ENERGY SCIENCE, ENGINEERING AND TECHNOLOGY

PIPELINE SAFETY

INCIDENT RESPONSE AND AUTOMATED VALVE USE

ENERGY SCIENCE, ENGINEERING AND TECHNOLOGY

Additional books in this series can be found on Nova's website under the Series tab.

Additional E-books in this series can be found on Nova's website under the E-book tab.

SAFETY AND RISK IN SOCIETY

Additional books in this series can be found on Nova's website under the Series tab.

Additional E-books in this series can be found on Nova's website under the E-book tab.

ENERGY SCIENCE, ENGINEERING AND TECHNOLOGY

PIPELINE SAFETY

INCIDENT RESPONSE AND AUTOMATED VALVE USE

MAEVE B. REAGAN
EDITOR

New York

For permission to use material from this book please contact us:
Telephone 631-231-7269; Fax 631-231-8175
Web Site: http://www.novapublishers.com

LIBRARY OF CONGRESS CATALOGING-IN-PUBLICATION DATA

ISBN: 978-1-62618-337-7

Published by Nova Science Publishers, Inc. † New York

CONTENTS

PREFACE

They crisscross underneath our cities and country sides, yet most of the time we are not even aware they are there. They deliver critical fuel that powers our homes, factories, and offices; and also transport the oil and gas that keep our cars, trucks, and planes operating. The nation's 2.5 million mile network of hazardous liquid and natural gas pipelines includes more than 400,000 miles of "transmission" pipelines, which transport products from processing facilities to communities and large-volume users. To minimize the risk of leaks and ruptures, the Pipeline and Hazardous Materials Safety Administration (PHMSA) requires pipeline operators to develop incident response plans. Pipeline operators with pipelines in highly populated and environmentally sensitive areas are also required to consider installing automated valves. This book examines the ability of transmission pipeline operators to respond to product release, with a focus on the advantages and disadvantages of installing automated valves in high-consequence areas and safeguarding the public.

Chapter 1 – The nation's 2.5 million mile network of hazardous liquid and natural gas pipelines includes more than 400,000 miles of "transmission" pipelines, which transport products from processing facilities to communities and large-volume users. To minimize the risk of leaks and ruptures, PHMSA requires pipeline operators to develop incident response plans. Pipeline operators with pipelines in highly populated and environmentally sensitive areas ("high-consequence areas") are also required to consider installing automated valves.

The Pipeline Safety, Regulatory Certainty, and Job Creation Act of 2011 directed GAO to examine the ability of transmission pipeline operators to respond to a product release. Accordingly, GAO examined (1) opportunities to improve the ability of transmission pipeline operators to respond to incidents and (2) the advantages and disadvantages of installing automated valves in high-consequence areas and ways that PHMSA can assist operators in deciding whether to install valves in these areas. GAO examined incident data; conducted a literature review; and interviewed selected operators, industry stakeholders, state pipeline safety offices, and PHMSA officials.

Chapter 2 – Statement of Senator John D. (Jay) Rockefeller IV, Chairman, U.S. Senate Committee on Commerce, Science, and Transportation.

Chapter 3 – Statement of Cynthia L. Quarterman, Administrator, Pipeline and Hazardous Materials Safety Administration.

Chapter 4 – Testimony of Deborah A.P. Hersman, Chairman, National Transportation Safety Board.

Chapter 5 – Testimony of Rick Kessler, President, The Pipeline Safety Trust.

In: Pipeline Safety
Editor: Maeve B. Reagan
ISBN: 978-1-62618-337-7

Chapter 1

PIPELINE SAFETY: BETTER DATA AND GUIDANCE NEEDED TO IMPROVE PIPELINE OPERATOR INCIDENT RESPONSE*

United States Government Accountability Office

WHY GAO DID THIS STUDY

The nation's 2.5 million mile network of hazardous liquid and natural gas pipelines includes more than 400,000 miles of "transmission" pipelines, which transport products from processing facilities to communities and large-volume users. To minimize the risk of leaks and ruptures, PHMSA requires pipeline operators to develop incident response plans. Pipeline operators with pipelines in highly populated and environmentally sensitive areas ("high-consequence areas") are also required to consider installing automated valves.

The Pipeline Safety, Regulatory Certainty, and Job Creation Act of 2011 directed GAO to examine the ability of transmission pipeline operators to respond to a product release. Accordingly, GAO examined (1) opportunities to improve the ability of transmission pipeline operators to respond to incidents and (2) the advantages and disadvantages of installing automated valves in high-consequence areas and ways that PHMSA can assist operators in deciding whether to install valves in these areas. GAO examined incident data;

* This is an edited, reformatted and augmented version of United States Government Accountability Office, Publication No. GAO-13-168, dated January 2013.

conducted a literature review; and interviewed selected operators, industry stakeholders, state pipeline safety offices, and PHMSA officials.

WHAT GAO RECOMMENDS

DOT should (1) improve incident response data and use these data to evaluate whether to implement a performance-based framework for incident response times and (2) share guidance and information on evaluation approaches to inform operators' decisions. DOT agreed to consider these recommendations.

WHAT GAO FOUND

The Department of Transportation's (DOT) Pipeline and Hazardous Materials Safety Administration (PHMSA) has an opportunity to improve the ability of pipeline operators to respond to incidents by developing a performance-based approach for incident response times. The ability of transmission pipeline operators to respond to incidents—such as leaks and ruptures—is affected by numerous variables, some of which are under operators' control. For example, the use of different valve types (manual valves or "automated" valves that can be closed automatically or remotely) and the location of response personnel can affect the amount of time it takes for operators to respond to incidents. Variables outside of operators' control, such as weather conditions, can also influence incident response time, which can range from minutes to days. GAO has previously reported that a performance-based approach—including goals and associated performance measures and targets—can allow those being regulated to determine the most appropriate way to achieve desired outcomes. In addition, several organizations in the pipeline industry have developed methods for quantitatively evaluating response times to incidents, including setting specific, measurable performance goals. While defining performance measures and targets for incident response can be challenging, PHMSA could move toward a performance-based approach by evaluating nationwide data to determine response times for different types of pipeline (based on location, operating pressure, and pipeline diameter, among other factors). However, PHMSA must first improve the data it collects on incident response times.

These data are not reliable both because operators are not required to fill out certain time-related fields in the reporting form and because operators told us they interpret these data fields in different ways. Reliable data would improve PHMSA's ability to measure incident response and assist the agency in exploring the feasibility of developing a performance-based approach for improving operator response to pipeline incidents.

The primary advantage of installing automated valves is that operators can respond quickly to isolate the affected pipeline segment and reduce the amount of product released; however, automated valves can have disadvantages, including the potential for accidental closures—which can lead to loss of service to customers or even cause a rupture—and monetary costs. Because the advantages and disadvantages of installing an automated valve are closely related to the specifics of the valve's location, it is appropriate to decide whether to install automated valves on a case-by-case basis. Several operators we spoke with have developed approaches to evaluate the advantages and disadvantages of installing automated valves. For example, some operators of hazardous liquid pipelines use spill-modeling software to estimate the amount of product release and extent of damage that would occur in the event of an incident. While PHMSA conducts a variety of information-sharing activities, the agency does not formally collect or share evaluation approaches used by operators to decide whether to install automated valves. Furthermore, not all operators we spoke with were aware of existing PHMSA guidance designed to assist operators in making these decisions. PHMSA could assist operators in making this decision by formally collecting and sharing evaluation approaches and ensuring operators are aware of existing guidance.

ABBREVIATIONS

DOT	Department of Transportation
NTSB	National Transportation Safety Board
PHMSA	Pipeline and Hazardous Materials Safety Administration

January 23, 2013
The Honorable John D. Rockefeller IV
Chairman
The Honorable Ranking Member
Committee on Commerce, Science, and Transportation
United States Senate

Honorable Fred Upton Chairman
The Honorable Henry Waxman
Ranking Member
Committee on Energy and Commerce
House of Representatives

The Honorable Bill Shuster
Chairman

The Honorable Nick J. Rahall
Ranking Member
Committee on Transportation and Infrastructure
House of Representatives

The United States has over 2.5 million miles of hazardous liquid and natural gas pipelines that transport approximately 65 percent of the energy we consume. These pipelines, which are largely regulated by the Department of Transportation's (DOT) Pipeline and Hazardous Materials Safety Administration (PHMSA), are relatively safe when compared with other modes of transporting hazardous goods (e.g., highway and rail). However, when pipelines leak or rupture the results can be devastating, including fatalities, injuries, and extensive property or environmental damage. Such an "incident" occurred in September 2010 in San Bruno, California, killing 8 people and damaging or destroying over 100 homes.[1] To minimize the risk of a pipeline incident, pipeline operators are required to develop leak detection methods and emergency response plans. Operators with pipelines in highly populated or environmentally sensitive areas (called "high-consequence areas") are subject to supplemental risk-based regulations under PHMSA's integrity management program.[2] Through this program, PHMSA requires that operators conduct a risk assessment to determine what additional measures to take to mitigate the consequences of pipeline failures. One mitigation measure operators can take based on the results of the risk assessment is to install automated valves, which in the event of an incident, close automatically or are closed remotely by operators in a control room.[3] Since 1971, the National Transportation Safety Board (NTSB) has made recommendations that DOT develop standards and requirements for automated valves. Following the San Bruno incident, NTSB recommended that DOT require natural gas pipeline operators install automated valves in all high-consequence areas.[4]

The Pipeline Safety, Regulatory Certainty, and Job Creation Act of 2011 mandated that GAO examine the ability of transmission pipeline[5] operators to respond to a hazardous liquid or natural gas release from an existing pipeline segment.[6] Accordingly, this report contains information on: (1) opportunities to improve the ability of transmission pipeline operators to respond to incidents, and (2) the advantages and disadvantages of installing automated valves in high-consequence areas and ways that PHMSA can assist operators in deciding whether to install valves in these areas.

To determine what opportunities exist to improve the ability of transmission pipeline operators to respond to incidents, we identified the variables that influence operators' incident response capabilities. To do so, we spoke with selected operators about their prior incidents and variables that influenced their ability to respond. Operators were selected based on criteria, including amount and types of pipeline owned in high-consequence areas[7] and geographic diversity. We also discussed prior incidents, incident response times, and federal oversight of the pipeline industry with officials from PHMSA, state pipeline safety offices, industry associations, and safety groups. Based on our discussions and review of prior incidents, we identified variables that influence operators' ability to respond to incidents. We also examined 2007 to 2011 PHMSA incident data, including data on:

- total number of incidents;
- type of incident (leak or rupture);
- type of pipeline where the incident occurred; and
- the dates and times when an incident occurred, the operator identified the incident, the operator's resources (personnel and equipment) arrived on site, and the operator shut down a pipeline or facility.

We assessed the reliability of data through discussions with PHMSA officials and select operators and determined that data elements related to numbers of incidents, types of releases, and types of pipeline where incidents occurred were reliable for the purpose of providing context. However, we determined that data elements related to response time were not sufficiently reliable for the purpose of conducting a detailed analysis of relationships between response time and other factors. Finally, we reviewed federal requirements, and industry and government performance standards related to emergency response within the pipeline industry.

To determine the advantages and disadvantages of installing automated valves in high-consequence areas and ways that PHMSA can assist operators in deciding whether to install these valves, we identified the key factors that should be used in deciding whether to install automated valves in high-consequence areas. To do so, we conducted a literature review of previous research dating back to 1995 and interviewed officials from industry associations and pipeline safety groups. In addition, we collected information from selected operators on their methods for deciding whether to install automated valves, as well as specific pipeline segments and valve locations on which they made such decisions. We also discussed the regulations with officials from PHMSA, state pipeline safety offices, and pipeline operators to determine what, if any, additional guidance would help operators apply the current regulations on installing automated valves. For further details on our scope and methodology, see appendix I.

We conducted this performance audit from March 2012 to January 2013 in accordance with generally accepted government auditing standards. Those standards require that we plan and perform the audit to obtain sufficient, appropriate evidence to provide a reasonable basis for our findings and conclusions based on our audit objectives. We believe that the evidence obtained provides a reasonable basis for our findings and conclusions based on our audit objectives.

BACKGROUND

Three main types of pipelines carry hazardous liquid[8] and natural gas from producing wells to end users (residences and businesses) and are managed by about 2,500 operators:

- *Gathering pipelines* collect hazardous liquid and natural gas from production areas and transport the products to processing facilities, which in turn refine and send the products to transmission pipelines. These pipelines tend to be located in rural areas but can also be located in urban areas. PHMSA estimates there are 200,000 miles of natural gas gathering pipelines and 30,000 to 40,000 miles of hazardous liquid gathering pipelines.
- *Transmission pipelines* carry hazardous liquid or natural gas, sometimes over hundreds of miles, to communities and large-volume users, such as factories. Transmission pipelines tend to have the largest diameters and operate at the highest pressures of any type of pipeline. PHMSA has estimated there are more than 400,000 miles of

hazardous liquid and natural gas transmission pipelines across the United States. (See fig. 1.)

- *Distribution pipelines* then split off from transmission pipelines to transport natural gas to end users—residential, commercial, and industrial customers. There are no hazardous liquid distribution pipelines. PHMSA has estimated there are roughly 2 million miles of natural gas distribution pipelines, most of which are intrastate pipelines.

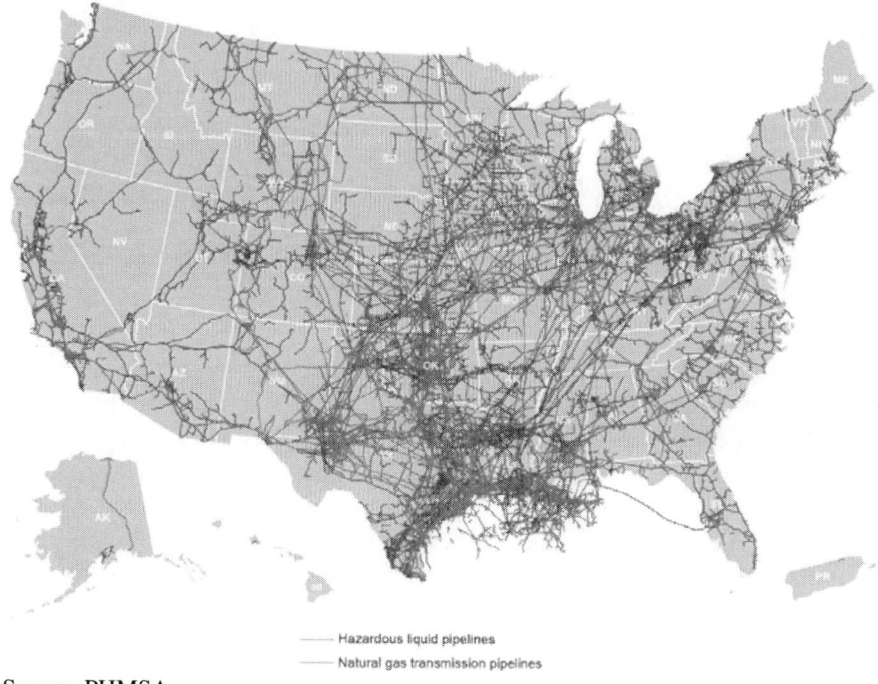

Source: PHMSA.

Figure 1. Transmission Pipeline across the United States, as of September 2012.

PHMSA administers the national regulatory program to ensure the safe transportation of hazardous liquid and natural gas by pipeline, including developing safety requirements that all pipeline operators regulated by PHMSA must meet.[9] In 2012, the agency's budget was $201 million, which was used, in part, to employ over 200 staff in its pipeline safety program. About half of the pipeline safety program staff inspects hazardous liquid and gas pipelines for compliance with safety regulations. Besides PHMSA, over

300 state inspectors help oversee pipelines and ensure safety. State and federal officials may also investigate specific pipeline incidents to determine the reason for the pipeline failure and to take enforcement actions, when necessary.[10]

PHMSA enforces two general sets of pipeline safety requirements. The first are minimum safety standards that cover specifications for the design, construction, testing, inspection, operation, and maintenance of pipelines. The second set of safety requirements are part of a supplemental risk-based regulatory program termed “integrity management.”[11] Under transmission pipeline integrity management programs, operators are required to systematically identify and mitigate risks to pipeline segments—discrete sections of the pipeline system separated by valves that can stop the flow of product—that are located in high-consequence areas where an incident would have greater consequences for public safety or the environment. To ensure operators comply with minimum safety standards and integrity management requirements, PHMSA conducts inspections in partnership with state pipeline safety agencies. Inspections may focus on specific pipeline segments or aspects of an operator’s safety program, or both. According to PHMSA, officials conduct an inspection for each operator at least once every 5 to 7 years, but may conduct additional inspections based on safety risk or at the discretion of PHMSA or state officials. PHMSA is authorized to take enforcement actions against operators, including issuing warning letters, notices of probable violation, notices of amendment, notices of proposed safety order, corrective action orders, and imposing civil penalties.[12]

Transporting hazardous liquids and natural gas by pipelines is associated with far fewer fatalities and injuries than other modes of transportation. From 2007 to 2011, there was an average of about 14 fatalities per year for all pipeline incidents reported to PHMSA, including an average of about 2 fatalities per year resulting from incidents on hazardous liquid and natural gas transmission pipelines. In comparison, in 2010, 3,675 fatalities resulted from incidents involving large trucks and 730 additional fatalities resulted from railroad incidents. Yet risks to pipelines exist, such as corrosion and third party excavation, which can damage a pipeline’s integrity and result in leaks and ruptures. A leak is a slow release of a product over a relatively small area. A rupture is a breach in the pipeline that may occur suddenly; the product may then ignite resulting in an explosion.[13] According to pipeline operators we met with, of the two types of pipeline incidents, leaks are more common but generally cause less damage. Ruptures are relatively rare but can have much higher consequences because of the damage that can be caused by an associated explosion.

According to PHMSA, industry, and state officials, responding to either a hazardous liquid or natural gas pipeline incident typically includes steps such as detecting that an incident has occurred, coordinating with emergency responders, and shutting down the affected pipeline segment. (See fig. 2.) Under PHMSA's minimum safety standards, operators are required to have a plan that covers these steps for all of their pipeline segments and to follow that plan during an incident. Officials from PHMSA and state pipeline safety offices perform relatively minor roles during an incident, as they rely on operators and emergency responders to take actions to mitigate the consequences of such events. Following an incident, operators must report incidents that meet certain thresholds— including incidents that involve a fatality or injury, excessive property damage or product release, or an emergency shutdown—to the federal National Response Center,[14] as well as conduct an investigation to identify the root cause and lessons learned. Federal and state authorities may also use their discretion to investigate some incidents, which can involve working with operators to determine the cause of the incident. If necessary, authorities will take steps to correct deficiencies in operator safety programs, including taking enforcement actions.

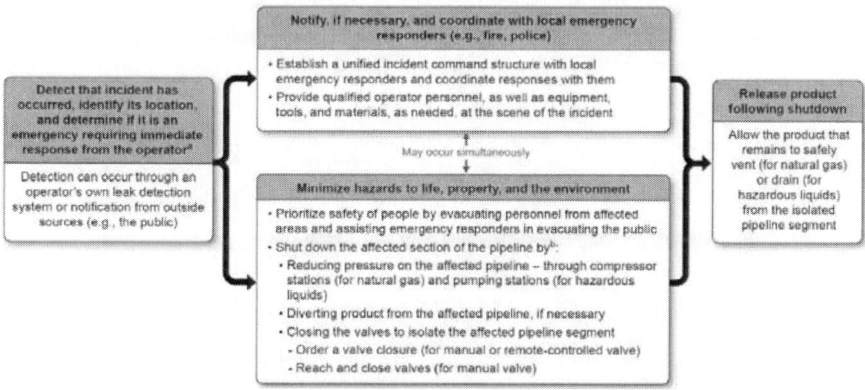

Source: GAO analysis, PHMSA, industry stakeholders, and pipeline operators.

[a] Some incidents, such as very small leaks, do not create a hazardous condition and may not require an immediate response from the operator. The operator can address these non-emergency incidents while maintaining normal operations.

[b] Emergency shutdown procedures can be initiated remotely by control room operators or through personnel in the field.

Figure 2. Steps Operators Take When Responding to Incidents.

While prior research shows that most of the fatalities and damage from an incident occur in the first few minutes following a pipeline rupture, operators can reduce some of the consequences by taking actions that include closing valves that are spaced along the pipeline to isolate segments. The amount of time it takes to close a valve depends upon the equipment installed on the pipeline. For example, valves with manual controls (referred to as "manual valves") require a person to arrive on site and either turn a wheel crank or activate a push-button actuator. Valves that can be closed without a person located at the valve location (referred to as "automated valves") include both remote-control valves, which can be closed via a command from a control room, and automatic-shutoff valves, which can close without human intervention based on sensor readings.[15,16] (See fig. 3.) Automated valves generally take less time to close than manual valves. PHMSA's minimum safety standards dictate the spacing of all valves, regardless of type of equipment installed to close them,[17] while integrity management regulations require that transmission pipeline operators conduct a risk assessment for high-consequence areas that includes the consideration of automated valves.[18]

Source: GAO.
A manual valve on the Northwest Pipeline system near Pocatello, Idaho.

Source: GAO.

A remotely-controlled valve on an Enterprise Products hazardous liquid pipeline system near Houston, Texas.

Figure 3. A Hand Wheel Used to Close a Manual Valve (Outlined in Red on Left) and an Actuator Used to Remotely Close an Automated Valve (Outlined in Red on Right).

PERFORMANCE-BASED APPROACH OFFERS OPPORTUNITY TO IMPROVE INCIDENT RESPONSE, BUT BETTER DATA ARE NEEDED

The ability of transmission pipeline operators to respond to incidents, such as leaks and ruptures, is affected by a number of variables—some of which are under operators' control—resulting in variances in response time; for a given incident, that time can range from minutes to days. Several states and industry organizations have developed performance-based requirements for operators to meet in responding to incidents. PHMSA has some performance-based requirements, but its current performance goal related to incident response is not well defined. More precise performance measures and targets could lead to improved response times and less damage from incidents in some cases.

However, PHMSA would need better data on incidents to determine the feasibility of such an approach.

Incident Response Time Depends on Multiple Variables, Some of Which Operators Can Control

According to PHMSA officials, pipeline safety officials, and industry stakeholders and operators, multiple variables—some controllable by transmission pipeline operators—can influence the ability of operators to respond quickly to an incident. Ensuring a quick response is important because according to pipeline operators and industry stakeholders, reducing the amount of time it takes to respond to an incident can also reduce the amount of property and environmental damage stemming from an incident and, in some cases, the number of fatalities and injuries. For example, several natural gas pipeline operators noted that a faster incident response time could reduce the amount of property damage from secondary fires (after an initial pipeline rupture) by allowing fire departments to extinguish the fires sooner. In addition, hazardous liquid pipeline operators told us that a faster incident response time could result in lower costs for environmental remediation efforts and less product lost. We identified five variables that can influence incident response time and that are within an operator's control:

- *Leak detection capabilities.* How quickly a leak is detected affects how soon an operator can initiate a response. Pipeline operators must perform a variety of leak detection activities to monitor their systems and identify leaks.[19] These activities commonly include periodic external monitoring, such as aerial patrols of the pipeline, as well as continuous internal monitoring, such as measuring the intake and outtake volumes or pressure flows on the pipeline. In addition, pipeline operators must conduct public awareness programs for those living near pipeline facilities about how to recognize, respond to, and report pipeline emergencies; these programs can influence how quickly an operator becomes aware of an incident. Attempting to confirm an incident can also affect response time. Pipeline operators may prefer to have two sources of information to confirm an incident, such as data from a pipeline sensor and a visual confirmation, especially if shutting down the system is a likely response to the incident. Natural gas pipeline operators in particular generally seek to confirm an incident before a shutdown, as shutdowns interrupt thc gas flow and can cut off service to their customers.

- *Location of qualified operator response personnel.* The proximity of the operator's response personnel to a facility or shutoff valve can affect the response time. Response personnel who have a greater distance to travel to the facility or valve site can take longer to establish an incident command center or to close manual valves. Along with proximity, incident response time depends on whether qualified operator response personnel—those who are trained and are authorized to take necessary action, such as closing manual valves—are dispatched.
- *Type of valves.* The type of valve an operator has installed on a pipeline segment can affect how quickly the segment can be isolated. Automated valves, which can be closed automatically or remotely, can shorten incident response time compared to manual valves, which require that personnel travel to the valve site and turn a wheel crank or activate a push-button actuator to close the valve. However, if affected valves happen to be located at or close to facilities where personnel are permanently stationed, the type of valve could be less critical in influencing incident response time.
- *Control room management.* Clear operating policies and shutdown protocols for control room personnel can influence response time to incidents.[20] For example, incident response time might be reduced if control room personnel have the authority to shut down a pipeline or facility if a leak is suspected, and are encouraged to do so. A few of the operators we met with told us that while in the past it was a common practice in the industry to avoid shutdowns unless absolutely necessary, the practice now for these operators is to shut down the line if there is any doubt about safety. An official from one natural gas pipeline operator told us that his company instructs control room personnel that they will not suffer repercussions from shutting down a line for safety reasons. Another official from a hazardous liquid pipeline operator told us that the authority to shut down is at the control room level and that even personnel in the field can make the call to shut down a line.
- *Relationships with local first responders.* Operators that have already established effective communications with local first responders—such as fire and police departments—may respond more quickly during emergencies.[21] For example, one natural gas pipeline operator

told us that during one incident, the local first responders had turned to the operator personnel for direction on how to respond to a rupture. As a result, the operator said that one of the lessons learned was that the company needed to conduct more emergency response exercises, such as mock drills, with the local first responders so the responders would know their roles and responsibilities.

We identified four other variables that influence a pipeline operator's ability to respond to an incident, but are beyond an operator's control:

- *Type of release.* The type of release—leak or rupture—can influence how quickly an operator responds to an incident. Leaks are generally a slow release of product over a small area, which can go undetected for long periods. Once a leak is detected, it can take additional time to confirm the exact location. Ruptures, which usually produce more significant changes in the external or internal conditions of the pipeline, are typically easier to detect and locate.
- *Time of day.* The time of day when an incident occurs can affect incident response time. The operator's response personnel may be delayed in reaching facilities in urban or suburban areas during peak traffic times. Conversely, if an incident occurs during the evening or on a weekend, the operator's response personnel could be able to reach the facility more quickly, because of lighter traffic. For example, one natural gas pipeline operator told us about an incident that occurred on a Saturday afternoon, which meant that traffic did not delay response personnel traveling to the scene.
- *Weather conditions.* Weather conditions can affect how quickly an operator can respond to an incident. For example, one natural gas pipeline operator described an incident caused by a hurricane's storm surge that pushed debris into the pipeline at a facility, and flooding prevented the response personnel from reaching the site for several days, during which time the pipe continued to leak gas. Winter conditions can also make it more difficult for the operator's response personnel to reach a facility or to access valve sites in remote areas. As another example, windy conditions can disperse natural gas and make it hard to detect a leak.
- *Other operators' pipeline in the same area.* If two or more operators own pipeline in a shared right of way,[22] determining whose system is affected can increase incident response time. Operators may delay

responding if they have not confirmed that the incident is on their pipeline. For example, one natural gas pipeline operator told us about an incident that took 2 days to repair because when their personnel first detected a leak, the personnel initially contacted another operator, whose line crossed over theirs, to make sure the leak was not the other operator's.

Operators we spoke with stated that the amount of time it takes to respond to an incident can depend on all of the variables listed above and can range from several minutes to days (see table 1).

A Performance-Based Approach Could Improve Incident Response Times

We and others have recommended that the federal government move toward performance-based regulatory approaches to allow those being regulated to determine the most appropriate way to achieve desired, measurable outcomes.[23] For example, Executive Order 13563 calls for improvements to the nation's regulatory system, including the use of the best, most innovative and least burdensome tools for achieving regulatory ends.[24] We have also previously reported on the benefits of a performance-based framework,[25] which helps agencies focus on achieving outcomes.[26] Such a framework should include: 1) national goals; 2) performance measures that are linked to those national goals; and 3) appropriate performance targets that promote accountability and allow organizations to track their progress towards goals. PHMSA has included these three elements of a performance-based framework in some aspects of its pipeline safety program, but not for incident response times. For example, PHMSA has set national goals intended to reduce the number of pipeline incidents involving fatality or major injury and the number of hazardous liquid pipeline spills with environmental consequences. Each of these national goals has associated performance measures (i.e., the number of such incidents) and specific targets (such as reducing the number of incidents involving a fatality or major injury from 39 to less than 28 per year by 2016) that allow PHMSA to track its progress toward the goals. However, while PHMSA has established a national goal for incident response times, it has not linked performance measures or targets to this goal.

Table 1. Examples of Response Times in Select Pipeline Incidents from 2009 to 2011

Incident response time	Description
1 minute	A rupture on a natural-gas transmission pipeline located underground in a sparsely populated area was caused when a construction company worker accidentally struck the pipeline, which then ignited and exploded. When the line broke, automatic-shutoff valves on either side of the rupture closed within one minute. Despite the fast valve closure, the explosion caused one fatality—the worker who struck the pipeline—and injured seven others. The affected pipeline segment was 20 miles long. Though the valveswere closed, there was enough gas remaining in the pipeline to fuel the fire for several hours. In addition to causing a fatality and injuries, the incident cost the operator an estimated $1 million, due primarily to the value of the lost product ($740,000), as well as damage to the pipeline ($288,000).
3 minutes	A rupture on a hazardous liquid transmission pipeline, located underground near a creek in a sparsely populated area, was caused when heavy rains shifted the land which broke the pipeline, releasing over 1,700 barrels of propane. The line break was immediately picked up by the operator's computer-based leak detection system, and operator personnel on site closed manual valves to isolate the segment within 3 minutes. Because propane is a highly volatile liquid, which turns to gas when released into the atmosphere, there was no soil or water contamination or environmental cleanup costs. The incident cost the operator an estimated $128,000, due primarily to the cost of repairs ($73,000) and value of lost product ($55,000).
8 minutes	During the night, unknown individuals operating construction equipment punctured a hazardous liquid transmission pipeline located underground in an environmentally sensitive area, causing 56 barrels of crude oil to leak into the soil. The puncture caused a drop in pressure that the control room operator detected in 2 minutes. Six minutes later, the control room operator shut down the pipeline and isolated the affected segment with remote-control valves. About two hours later, the operator's response personnel arrived on site. The incident cost the operator an estimated $1.3 million, due primarily to its environmental remediation efforts ($1 million) and emergency response ($250,000).
2 hours	A crack on an above-ground portion of a hazardous liquid pipeline, located in a populated area, caused 120 barrels of crude oil to spray into the air. About 15 minutes after the incident started, a local resident reported to the fire department that crude oil was spraying into the air at a pipeline station. The fire department went to the incident site and, about 30 minutes after the initial call, notified the pipeline operator of a broken oil pipeline. About 20 minutes after receiving the fire department's call, the control room began shutting down the pipeline system and isolating the affected segment by ordering the closure of the upstream valve. Approximately 50 minutes later—about 2 hours after the incident started—response personnel arrived on site and manually closed the valve, which stopped the leak. The incident cost the operator an estimated $183,000, due primarily to its emergency response ($118,000) and environmental remediation efforts ($61,000).

Incident response time	Description
7 days	A natural gas transmission pipeline, located underground in a sparsely populated area, developed a small leak as the result of a construction defect. The operator did not discover the leak on the pipeline for almost a week following initial reports due to the size of the leak in combination with wind gusts in the area that dissipated the escaping natural gas, reducing the common signs of a gas leak, such as the smell and damage to vegetation. Once the operator detected the leak during routine, periodic external monitoring of the pipeline, it took over a day to identify its exact location. The incident cost the operator an estimated $128,000 in repairs ($106,000) and lost product ($22,000).

Source: GAO presentation of information obtained during interviews with pipeline operators.

Specifically, PHMSA directs operators to respond to certain incidents–emergencies that require an immediate response–in a "prompt and effective" manner,[27] but neither PHMSA's regulations nor its guidance describe ways to measure progress toward meeting this goal. Without a performance measure and target for a prompt and effective incident response, PHMSA cannot quantitatively determine whether an operator meets this goal. PHMSA officials told us that because each incident presents unique circumstances, its inspectors must determine whether an operator's incident response was prompt and effective on a case-by-case basis. According to PHMSA, in making this determination, inspectors must use their professional judgment to balance any challenges the operator faced in responding with the operator's obligation to the public's safety.

Other organizations in the pipeline industry, including some state regulatory agencies, have developed methods for measuring the performance of operators responding to incidents by using specific incident response times. According to the National Association of Pipeline Safety Representatives, several state pipeline safety offices have initiatives that require natural gas pipeline operators to respond within a specified time frame to reports of pipeline leaks. For example, the New Hampshire Public Utilities Commission has established incident response time standards—ranging from 30 to 60 minutes, with performance targets—for natural gas distribution companies to meet when responding to reports of a leak.[28] In addition, members of the Interstate Natural Gas Association of America have committed to achieving a 1-hour incident response time for large diameter (greater than 12 inches) natural gas pipelines in highly populated areas.[29] To meet this goal, operators are planning changes to their systems, such as relocating response personnel and automating over 1,800 valves throughout the United States.

According to PHMSA officials, pipeline incidents often have unique characteristics, so developing a performance measure and associated target for incident response time similar to those used by other pipeline organizations would be difficult. In particular, it would be challenging to establish a performance measure using incident response time in a way that would always lead to the desired outcome of a prompt and effective response. Officials stated that the intention behind requiring operators to respond promptly and effectively is to make the area safe as quickly as possible. In some instances, an operator can accomplish this outcome in the time it takes to close valves and isolate pipeline segments, while in other instances, an operator might need to completely vent or drain the product from the pipeline. Likewise, it would be difficult to identify a specific target for incident response time, as pipeline operators likely should respond to some incidents more quickly than others. For example, industry officials noted that while most fatalities and injuries caused by a pipeline explosion occur in the initial blast, a faster incident response time could help reduce fatalities and injuries in cases where there are sites nearby whose occupants have limited mobility (e.g., prisons, hospitals). In these situations, operators told us they want to ensure their incident response time is faster than for more remote locations where an explosion would have less of an impact on people, property, and the environment.

Although defining performance measures and targets for incident response can be challenging, one way for PHMSA to move toward a more quantifiable, performance-based approach would be to develop strategies to improve incident response based on nationwide data. For example, performing an analysis of nationwide incident data—similar to PHMSA's current analyses of fatality and injury data—could help PHMSA determine response times for different types of pipelines (based on characteristics such as location, operating pressure, and diameter); identify trends; and develop strategies to improve incident response. Furthermore, as part of this analysis of response times for various types of pipelines, PHMSA could explore the feasibility of integrating incident response performance measures and targets for individual pipelines into its integrity management program. For example, PHMSA might identify performance measures that are appropriate for various types of pipelines and allow operators to determine which measures and targets best apply to their individual pipeline segments, based on the characteristics of those segments. Such an approach would be consistent with our prior work on performance measurement, as it would allow operators the flexibility to meet response time targets in several ways, including changes to their leak detection methods, moving personnel closer to the valve location, or installing automated valves. PHMSA would then review an operator's selection of measures and targets as part of ongoing integrity management inspections; this process is similar to how inspectors review other provisions in the integrity management program.

PHMSA Data on Incident Response Time Are Limited

PHMSA would need reliable national data to implement a performance-based framework for incident response times to ensure operators are responding in a prompt and effective manner.[30] However, the data currently collected by PHMSA do not enable them to accurately determine incident response times for all recent incidents for two reasons: 1) operators are not required to fill out certain time-related fields in the PHMSA incident-reporting form and 2) when operators do provide these data, they are interpreting the intended content of the data fields in different ways. Specifically, PHMSA requires operators to report the date and time when the incident occurred. Operators are not required to report the dates and times when:

- the operator identified the incident;
- the operator's resources (personnel or equipment) arrived on site; and
- the operator shut down and restarted a pipeline or facility.

As a result, our analysis determined that hazardous liquid pipeline operators did not report the date and time for two of these variables— when the incident was identified and when operator resources arrived on site—for 26 percent (178 out of 674) of incidents that occurred in 2010 and 2011. Also, these operators did not identify whether a shutdown took place in 16 percent (108 out 674) of incidents over the same time period.[31] In comparison, natural gas pipeline operators reported more complete data; these operators did not report data for when the operator identified the incident and resources arrived on site in only 3 percent (6 out of 191) of incidents that occurred in 2010 and 2011. Also, these operators did not identify whether a shutdown took place in only about 2 percent (3 out of 191) of incidents over the same period. PHMSA officials told us that because they have not used the time-related data to identify safety trends, the omissions have not been a problem for them, although in the future they may decide to make some of these data fields mandatory.

In addition to omitting certain incident data fields, several officials from pipeline operators told us that they interpret what to include in the time-related, incident data fields differently. For example, according to one official from a natural gas operator, some operators interpret the time when an operator identified the incident as the time when operator personnel first received a call about a potential leak, while others may interpret the time when an operator identified an incident as the time when operator personnel received an on-site confirmation of a leak. These differing interpretations occur even though guidance on PHMSA's website instructs operators how to complete the reporting forms, including the time-related data fields.[32]

IMPROVED INFORMATION SHARING ABOUT EVALUATING AUTOMATED VALVE ADVANTAGES AND DISADVANTAGES COULD INFORM OPERATORS' DECISIONS

The primary advantage of installing automated valves is reducing the time to shut down and isolate a pipeline segment after a leak or rupture occurs, while disadvantages include the potential for accidental closures and monetary cost. Because these advantages and disadvantages vary among valve locations, operators should make decisions about whether to install automated valves—as opposed to other safety measures—on a case-by-case basis. PHMSA has several opportunities to assist operators in making these evaluations, including communicating guidance and sharing information on some methods operators use to make these decisions.

Improved Response Times, Potential for Accidental Closures, and Costs Should be Evaluated on a Case-by-Case Basis

Research and industry stakeholders indicate that the primary advantage of installing automated valves is related to the time it takes to respond to an incident.

Although automated valves cannot mitigate the fatalities, injuries, and damage that occur in the initial blast, quickly isolating the pipeline segment through automated valves can significantly reduce subsequent damage by reducing the amount of hazardous liquid and natural gas released. For example, NTSB found that automated valves would have reduced the amount of time taken to stop the flow of natural gas in the San Bruno incident and, therefore, reduced the severity of property damage and life-threatening risks to residents and emergency responders.[33]

According to research and industry stakeholders, automated valves will only decrease the number of fatalities and injuries in those cases when people cannot easily evacuate the area, such as cases involving hospital patients or prison inmates.

Table 2. Advantages and Disadvantages of Installing Automated Valves on Pipelines

Advantages	Improved response time	• Can reduce injuries and fatalities for some locations, such as hospitals or prisons, where people cannot evacuate quickly. • Can reduce the amount of damage by limiting the amount of fuel for secondary fire(s) and environmental cleanup. • Can allow operator personnel and emergency responders to access the affected segment more quickly and safely. • Can reduce the potential monetary cost of an incident for the operator by limiting the amount of product lost.
Disadvantages	Accidental closures	• For natural gas pipelines, accidental closures can result in the loss of service to utilities and critical customers (e.g., winter-time outages can leave people without heat). • For hazardous liquid pipelines, accidental closures can cause an incident, whena valve closes and the subsequent pressure buildup causes the pipeline to rupture.
	Monetary costs	• Requires operators to purchase equipment, including devices to remotely communicate or sense pressure drops, actuators to close the valve, and power sources for this new equipment. • Requires operators to take on installation costs, which can involve temporarily shutting down the pipeline, purging the product from the pipeline, and pulling product from the market. Operators may also have costs related to accessing the valve location (e.g., right of way, permitting, and physical space to install thenew equipment) and updating their leak detection technologies. •May require operators to incur additional recurring costs to train staff, maintain the valves, increase security, and conduct inspections of the new valve.

Source: GAO analysis of research and industry stakeholder opinions.

Research and industry stakeholders identified several disadvantages operators should consider when determining whether to install automated valves, related to potential accidental closures and the monetary costs of purchasing and installing the equipment. Specifically, automated valves can lead to accidental closures, which can have severe, unintended consequences, including loss of service to residences and businesses. For example, according to a pipeline operator, an accidental closure on a natural gas pipeline in New Jersey resulted in significant disruption and downstream curtailments to customers in New York City during high winter demand. In addition, the monetary costs of installing automated valves can range from tens of thousands to a million dollars per valve, which may be significant expenditures for some pipeline operators.[34] (See table 2.)

Research and industry stakeholders also indicate the importance of determining whether to install valves on a case-by-case basis because the advantages and disadvantages can vary considerably based on factors specific to a unique valve location. These sources indicated that the location of the valve, existing shutdown capabilities, proximity of personnel to the valve location, the likelihood of an ignition, type of product being transported, operating pressure, topography and pipeline diameter, among others, all play a role in determining the extent to which an automated valve would be advantageous.

Operators Have Developed Approaches to Evaluate Advantages and Disadvantages of Installing Automated Valves

Operators we met with are using a variety of methods for determining whether to install automated valves. One of the eight operators we met with had decided to install automatic-shutoff valves across its pipeline system, regardless of risk, to eliminate the need for control room staff to make judgment calls on whether or not to close valves to isolate pipeline segments. However, seven of the eight operators we met with developed their own risk-based approach for considering potential advantages and disadvantages when making these decisions on a case-by-case basis.

For example, two natural gas pipeline operators told us that they applied a decision tree analysis to all pipeline segments in highly populated and frequented areas.

They used the decision tree to guide a variety of yesor-no questions on whether installing an automated valve would improve response time to less than an hour and provide advantages for locations where people might have difficulty evacuating quickly in the event of a pipeline incident. Other operators said they used computer-based spill modeling to determine whether the amount of product release would be significantly reduced by installing an automated valve.

These seven operators told us that their approaches for making decisions about whether to install automated valves considered the advantages and the disadvantages we identified above.[35]

- *Improved response time.* Most operators we spoke with considered whether automated valves would lead to a faster response time. For example, the primary criterion used by two of the natural gas pipeline operators was the amount of time it would take to shut down the pipeline and isolate the segment and population along the segment. In one instance, an operator decided to install a remote-control valve in a location that would take pipeline personnel 2.5 hours to reach and 30 minutes more to close the valve. Installing the automated valve is expected to reduce the total response time to under an hour, including detecting the incident and making the decision to isolate the pipeline segment. In addition, several hazardous liquid pipeline operators used spill modeling to determine whether an automated valve would result in a reduced amount of damage from product release at individual locations. This spill modeling typically considered topography, operating pressure, and placement of existing valves. For example, one hazardous liquid pipeline operator used spill modeling to make the decision to install a remote-control valve on a pipeline segment with a large elevation change after evaluating the spill volume reduction.
- *Accidental closures.* Operators indicated that installing automated valves, especially automatic-shutoff valves, could have unintended consequences, which they considered as part of their decisions to install automated valves. For example, two natural gas pipeline operators considered whether there is the potential for accidentally cutting off service when assessing individual locations for the possible installation of an automatic-shutoff valve. As noted, one natural gas pipeline operator has made the decision to install automatic-shutoff valves across its pipeline system. The operator stated that in the past, there were concerns with relying on automatic-shutoff valves because of the possibility for accidental closures, but the operator believes it

has developed a process that effectively adapts to pressure and flow change and minimizes or eliminates the risk of the valve accidentally closing. Other natural gas pipeline operators stated that relying on pressure sensing systems can be dangerous because "tuning" the pressure activation in an effort to avoid accidental closures can result in situations where the valve will not automatically close during an actual emergency. For hazardous liquid, all operators we spoke with stated that they either do not consider or do not typically install automatic-shutoff valves because an accidental closure has the potential to lead to an incident. Specifically, operators stated that an unexpected valve closure can result in decompression waves in the pipeline system, which might cause the pipeline to rupture if operators cannot reduce the flow of product promptly.

- *Monetary costs.* According to operators and other industry stakeholders, considering monetary costs is important when making decisions to install automated valves because resources spent for this purpose can take away from other pipeline safety efforts. Specifically, operators and industry stakeholders told us they often would rather focus their resources on incident prevention to minimize the risk of an incident instead of focusing resources on incident response. PHMSA stated that it generally supports the idea that pipeline operators should be given flexibility to target compliance dollars where they will have the most safety benefit when it is possible to do so. Operators we spoke with stated that they considered costs associated with purchasing and installing equipment. For example, four operators indicated that they will consider the costs related to communications equipment when determining whether to install automated valves. In addition, three operators stated that decisions to install automated valves are affected by whether the operator has or can gain access to the pipeline right of way. Other cost considerations mentioned by at least one operator included local construction costs and possible changes to leak detection systems. Finally, two natural gas pipeline operators stated that monetary cost plays a role in determining what steps they plan to take to meet a one-hour response time goal for pipelines in highly populated areas. For example, the operator might choose to move personnel closer to valves rather than installing automated valves, if that is the more cost-effective option.[36]

Opportunities Exist for PHMSA to Better Communicate Guidance and Share Best Practices Operators Use to Determine Whether to Install Automated Valves

PHMSA has developed guidance to help operators understand current regulations[37] on what operators must consider when deciding to install automated valves, but not all operators are aware of the guidance. PHMSA includes on its primary website two types of guidance that can be useful for operators in determining whether to install automated valves on transmission pipelines. First, PHMSA has developed inspection protocols for both the hazardous liquid and natural gas integrity management program. Second, PHMSA has developed guidance on the enforcement actions inspectors will take—such as a notice of proposed violation and warning letter, among others—should PHMSA discover a violation. Both of these pieces of guidance provide additional detail—not included in regulation—on the steps operators might take in considering whether to install automated valves. For example, PHMSA's inspection protocol for natural gas operators describes several studies on the generic costs and benefits of automated valves and indicates that operators may use this research as long as they document the reasons why the study is applicable to the specific pipeline segment. However, operators we spoke with were unaware of existing guidance to varying degrees. Specifically, of the eight operators we met with, three were unaware of both the inspection and enforcement guidance, and the remaining five operators were unaware of the enforcement guidance. Operators we spoke with, including those that were unaware of the guidance, told us that having this information would be helpful in making decisions to install automated valves. According to PHMSA, the agency provides this guidance to operators to ensure operators follow it as they make decisions on whether to install automated valves, but does not re-distribute the guidance at regular intervals (e.g., annually).

According to PHMSA, inspectors see examples of how operators make decisions to install automated valves during integrity management inspections, but the inspectors do not formally collect this information or share it with other operators. Current regulations give operators a large degree of flexibility in making decisions in deciding to install automated valves. As mentioned earlier, we spoke with operators that are using a variety of risk-based methods for making decisions about automated valves. For example, some used basic yes-or-no criteria, while others applied commercially available computer software to model potential incident outcomes. According to PHMSA, officials do not formally share what they view as good methods for determining whether to install automated valves. Officials stated they do not believe it is appropriate for PHMSA to publicly share decision-making approaches from a single operator, as doing so might be seen as an endorsement of that approach. However, according to PHMSA, its inspectors may informally discuss methods used by operators for making decisions to install automated valves and suggest these approaches to other operators during inspections. While the operators we spoke with represent roughly 18 percent of the overall hazardous liquid and natural gas transmission pipelines in high-consequence areas in 2010, there are over 650 additional pipeline operators we did not speak with that may be using other methods for determining whether to install automated valves. As such, we believe that both operators and inspectors could benefit from exposure to some of the methods used by other operators to make decisions on whether to install automated valves. We have previously reported on the value of organizations reporting and sharing information and recommended that PHMSA develop methods to share information on practices that can help ensure pipeline safety.[38] PHMSA already conducts a variety of information-sharing activities that could be used to ensure operators are aware of both existing guidance and of approaches used by other operators for making decisions to install valves. While, according to PHMSA officials, the agency will not endorse a particular operator's approach or practice, it can and does facilitate the exchange of information among operators and other stakeholders. For example, PHMSA issues advisory alerts in the Federal Register on emerging safety issues, including identified mechanical defects on pipelines, incidents that occurred under special circumstances, and reminders to correctly implement safety programs (e.g., drug and alcohol screening). In addition, PHMSA administers a website different from its primary website that, according to officials, is intended to ensure communication with pipeline safety stakeholders, including the public, emergency officials, pipeline safety advocates, regulators, and pipeline operators.[39] PHMSA also periodically conducts public workshops with pipeline stakeholders on a wide variety of topics, including one in March 2012 on automated valves.

CONCLUSION

While PHMSA currently requires operators to respond to incidents in a "prompt and effective manner," the agency does not define these terms or collect reliable data on incident response times to evaluate an operator's ability to respond to incidents. A more specific response time goal may not be appropriate for all pipelines. However, some organizations in the pipeline industry believe that such a performance-based goal can allow operators to identify actions that could improve their ability to respond to incidents in a timelier manner, and are taking steps to implement a performance-based approach. A performance-based goal that is more specific than "prompt and effective" could allow operators to examine the numerous variables under their control within the context of an established time frame to understand their current ability to respond and identify the most effective changes to improve response times, if needed, on individual pipeline segments. Reliable data would improve PHMSA's ability to measure incident response and assist the agency in exploring the feasibility of developing a performance-based approach for improving operator response to pipeline incidents.

One of the methods operators could choose to meet a performance-based approach to incident response is installing automated valves, a measure some operators are already taking to reduce risk. Given the different characteristics among valve locations, it is important for operators to carefully weigh the potential for improved incident response times against any disadvantages, such as the potential for accidental closure and monetary costs, in deciding whether to install automated valves as opposed to other safety measures. However, not all operators we spoke with were aware of existing PHMSA guidance and PHMSA does not formally collect or share evaluation approaches used by other operators to make decisions about whether to install automated valves. Such information could assist operators in evaluating the advantages and disadvantages of these valves and help them determine whether automated valves are the best option for meeting a performance-based incident response goal.

Recommendations for Executive Action

We recommend that the Secretary of Transportation direct the PHMSA Administrator to take the following two actions:

- To improve operators' incident response times, improve the reliability of incident response data and use these data to evaluate whether to implement a performance-based framework for incident response times.

- To assist operators in determining whether to install automated valves, use PHMSA's existing information-sharing mechanisms to alert all pipeline operators of inspection and enforcement guidance that provides additional information on how to interpret regulations on automated valves, and to share approaches used by operators for making decisions on whether to install automated valves.

AGENCY COMMENTS

We provided the Department of Transportation with a draft of this report for review and comment. The department had no comments and agreed to consider our recommendations.

Susan Fleming
Director, Physical Infrastructure

APPENDIX I: OBJECTIVES, SCOPE, AND METHODOLOGY

The objectives of our review were to determine (1) the opportunities that exist to improve the ability of transmission pipeline operators to respond to incidents and (2) the advantages and disadvantages of installing automated valves in high-consequence areas and ways that the Pipeline and Hazardous Materials and Safety Administration (PHMSA) can assist operators in deciding whether to install valves in these areas.

To address our objectives, we reviewed regulations, National Transportation Safety Board (NTSB) incident reports, and PHMSA guidance and data on enforcement actions, pipeline operators, and incidents, related to onshore natural gas transmission and hazardous liquid pipelines. We also attended industry conferences and interviewed officials at PHMSA headquarters and regional offices (Eastern, Southwestern, and Western), state pipeline safety agencies, pipeline safety groups, and industry associations. Specifically, we interviewed officials from the American Gas Association, American Petroleum Institute, Arizona Office of Pipeline Safety, Association of Oil Pipelines, Interstate Natural Gas Association of America, National Association of Pipeline Safety Representatives, NTSB, Pipeline Research

Council International, Public Utilities Commission of Ohio, and West Virginia Public Service Commission.

To address both objectives, we also conducted case studies on eight hazardous liquid and natural gas pipeline operators. We selected these operators based on our review of PHMSA data on the operators' onshore pipeline mileage, product type and prior incidents, recommendations from industry associations and PHMSA, and to ensure geographic diversity. We selected six hazardous liquid and natural gas pipeline operators with a large amount of pipeline miles in high-consequence areas that also reported recent incidents (i.e., one or more incident(s) reported from 2007 through 2011) with a range of characteristics, such as: affected a high-consequence area; resulted in an ignition/explosion; or involved an automated valve. We also selected one natural gas pipeline operator and one hazardous liquid pipeline operator with a small number of pipeline miles in high-consequence areas, to obtain the perspective of smaller pipeline operators.[1] Specifically, we interviewed officials from:

- Belle Fourche Pipelines (Casper, Wyoming)—Hazardous Liquids;
- Buckeye Partners (Breinigsville, Pennsylvania)—Hazardous Liquids;
- Enterprise Products (Houston, Texas)—Hazardous Liquids and Natural Gas;
- Granite State Gas Transmission (Portsmouth, New Hampshire)—Natural Gas;
- Kinder Morgan-Natural Gas Pipeline Company (Houston, Texas)—Natural Gas;
- Phillips 66 (Houston, Texas)—Hazardous Liquids;
- Northwest Pipeline GP (Salt Lake City, Utah)—Natural Gas; and
- Williams-Transco (Houston, Texas)—Natural Gas.

To determine what opportunities exist to improve the ability of transmission pipeline operators to respond to incidents, we identified several factors that influence pipeline operators' incident response capabilities. To do so, we discussed prior incidents, incident response times, and federal oversight of the pipeline industry with officials from PHMSA, state pipeline safety offices, industry associations, and safety groups. We also spoke with operators about their prior incidents and the factors that influenced their ability to respond. We also examined 2007 to 2011 PHMSA incident data, including data on total number of incidents, type of incident (leak or rupture), type of pipeline where the incident occurred, and the date and time when: an incident occurred; an operator identified the incident; operator resources (personnel and equipment) arrived on site; and an operator shut down a pipeline or facility. We assessed the reliability of these data through discussions with PHMSA officials and selected operators. We determined that data elements related to numbers of incidents, types of releases, and types of pipeline where incidents occurred were reliable for the purpose of providing context, but that data elements related to response time were not sufficiently reliable for the purpose of conducting a detailed analysis of relationships between response time and other factors. We also reviewed federal requirements, prior GAO reports, and industry and government performance standards related to emergency response within the pipeline industry.

To determine the advantages and disadvantages of installing automated valves in high-consequence areas and the ways that PHMSA can assist operators in deciding whether to install these valves, we identified the key factors that should be used in deciding whether to install automated valves in high-consequence areas. We used two categories of sources to identify the key factors:

1) *Literature review.* We conducted a literature review of previous research on pipeline incidents. Specifically, we used online research software to search through databases of scholarly and peer-reviewed materials—including articles, journals, reports, studies, and conferences dating back to 1995—which identified over 200 sources.
2) *Interviews with industry stakeholders.* During our interviews with officials from industry associations and pipeline safety groups, we discussed the advantages and disadvantages of installing these valves.

To ensure that the literature review included just those documents that were relevant to our purpose, two analysts independently reviewed abstracts from the 200 sources identified to determine whether they were within the scope of our review. Each source had to meet specific criteria, including mentioning automated valves, pipeline incidents, and operator emergency response. We excluded sources that were overly technical for the purposes of our review. To ensure these analysts were making similar judgments, they separately examined a random sampling of each other's sources. The analysts then added sources suggested by industry stakeholders during our interviews and reviewed them using the same criteria. After excluding documents that were not publicly available, one analyst reviewed these sources to identify advantages and disadvantages operators should consider when making decisions to install automated valves. A second analyst reviewed the analysis and performed a spot check on identified advantages and disadvantages. Specifically, the second analyst picked four of the sources at random to review and compared the advantages and disadvantages he identified to those of the first analyst.

As part of our case studies, we discussed these advantages and disadvantages with operators. We also collected information from operators on their methods for deciding whether to install automated valves, as well as specific pipeline segments and valve locations where operators made such decisions (see app. II). We contacted vendors (manufacturers and installers) of automated valves to identify the range of costs for purchasing and installing these valves.

We also discussed the regulations with officials from PHMSA headquarters and regional offices, state pipeline safety offices, and pipeline operators to determine what, if any, additional guidance would help operators apply the current regulations on installing automated valves.

We conducted this performance audit from March 2012 to January 2013 in accordance with generally accepted government auditing standards.

Those standards require that we plan and perform the audit to obtain sufficient, appropriate evidence to provide a reasonable basis for our findings and conclusions based on our audit objectives. We believe that the evidence obtained provides a reasonable basis for our findings and conclusions based on our audit objectives.

APPENDIX II: HOW SELECT OPERATORS DETERMINED WHETHER TO INSTALL AUTOMATED VALVES

We conducted site visits to eight hazardous liquid and natural gas pipeline operators with different amounts of pipeline miles in or affecting high-consequence areas.[1]

Seven of the eight operators we visited told us they use approaches that consider both the advantages and disadvantages of installing automated valves on a case-by-case basis as opposed to other safety measures; the eighth operator stated that it follows a corporate strategy of installing automated valves in all high-consequence areas. A brief description of the approach used by each of the eight operators, based on our discussions with them, follows.

Pipeline operator: Belle Fourche Product type: Hazardous liquid

Number of pipeline miles: 460 (total); 135 (could affect high-consequence areas)

Decision-making approach: The operator assesses each pipeline segment[2] using spill-modeling software to determine the amount of product release and extent of damage that would occur in the event of an incident. The software considers flow rates, pressure, terrain, product type, and whether the segment is located over land or a waterway. Monetary costs are considered as part of the decision-making process, including the cost of installing communications equipment and gaining access to the valve location when the operator does not own the right of way. The operator stated that installing a remote-control valve costs between $100,000 and $500,000. Automatic-shutoff valves are not considered as the operator believes an accidental closure could lead to pipeline ruptures.

Results to date: According to Belle Fourche officials, this approach has not resulted in any decisions to install automated valves because the advantages have not outweighed the disadvantages on any of the pipeline segments assessed.[3]

Pipeline operator: Buckeye Partners

Product type: Hazardous liquid

Number of pipeline miles: 6,400 (total); 4,179 (could affect high-consequence areas)

Decision-making approach: The operator assesses each pipeline segment using spill-modeling software to determine the amount of product release and extent of damage that would occur in the event of an incident. The operator considers installation of an automated valve when this modeling shows such a valve would 1) reduce the size of the incident by 50 percent or more and 2) significantly reduce the consequences of an incident. The operator conducts additional analysis to determine the location where the automated valve would lead to the largest reduction in spill volume and overall consequences of an incident. Monetary costs are considered as part of the decision-making process, including costs for gaining access to pipeline when the operator does already not own the right of way. The operator stated that installing a remote-control valve costs between $35,000 and $325,000. Automatic-shutoff valves are considered, but not typically installed, as the operator believes an accidental closure could lead to a pipeline rupture.

Results to date: According to Buckeye Partners officials, this approach has resulted in additional analysis of the possible installation of 25 remote-control valves along 75 pipeline segments assessed.

Pipeline operator: Phillips 66 Product type: Hazardous liquid

Number of pipeline miles: 11,290 (total); 3,851 (could affect high-consequence areas)

Decision-making approach: The operator assesses every 100 feet of pipeline (which covers all pipeline segments) using spill-modeling software to determine the amount of product release and extent of damage that would occur in the event of a complete rupture. The operator also uses a relative consequence index for individual pipeline segments that considers the impact to high-consequence areas. Automated valve projects are further evaluated if 1) the potential drain volume is greater than 1,000 barrels, 2) the pipeline segment exceeds a certain threshold on the consequence index, or 3) the existing automated valves are greater than 7.5 miles apart. Monetary costs are considered as part of the decision-making process, including the cost of installing communications equipment, access to power, gaining access to the valve's location when the operator does not own the right of way, and local construction costs. The operator stated that installing an automated valve costs between $250,000 and $500,000. Automatic-shutoff valves are not considered as the operator believes an accidental closure could lead to pipeline ruptures.

Results to date: According to the Phillips 66 officials, this approach has resulted in decisions to install 71 automated valves in the 508 high-consequence area locations assessed.

Pipeline operator: Enterprise Products Product type: Hazardous liquid and natural gas

Number of pipeline miles: 23,012 (total); 8,783 (could affect or in high-consequence areas)

Decision-making approach: The operator assesses each pipeline segment using spill-modeling software to determine the amount of product release and extent of damage that would occur in the event of an incident. The software considers factors such as topography and the placement of existing valves. The operator also uses a risk algorithm to identify threats to individual pipeline segments. The operator told us that it does not have specific criteria for guiding decisions to install automated valves; rather, officials make judgment calls based on the results of spill modeling and the application of the risk algorithm. Monetary costs are considered as part of the decision-making process, including the cost of installing communications equipment and the amount of necessary infrastructure work. The operator stated that installing a remote-control valve costs between $250,000 and $500,000. Pipelines carrying gas or highly volatile liquids—which are in gas form when released into the atmosphere—are excluded from consideration, according to the operator, because industry studies have shown that automated valves do not significantly improve incident outcomes for these product types.

Results to date: According to Enterprise Products officials, this approach has not resulted in any decisions to install automated valves because the advantages have not outweighed the disadvantages on any of the pipeline segments assessed.

Pipeline operator: Granite State Gas Transmission

Product type: Natural gas

Number of pipeline miles: 86 (total); 11 (high-consequence areas)

Decision-making approach: The operator assesses individual pipeline segments in high-consequence areas using risk analysis software that considers the operator's response time to an incident, population in the area, and pipeline diameter, among other variables. Monetary costs are considered as part of the decision-making process, including the cost of installing communications equipment and costs to change or improve the existing leak detection system. The operator stated that installing an automated valve costs between $40,000 and $50,000. Automatic-shutoff valves are not considered, as officials believe that they could lead to unintended consequences, such as accidental closures.

Results to date: According to Granite State Gas Transmission officials, this approach has resulted in decisions to install remote-control valves in 30 of the 30 locations assessed.

Pipeline operator: Kinder Morgan-Natural Gas Pipeline Company of America (NGPL)

Product type: Natural gas

Number of pipeline miles: 9,800 (total); 569 (high-consequence areas)

Decision-making approach: The operator follows a long-term corporate risk management strategy for NGPL, developed in the 1960s, that calls for installing automatic-shutoff valves across its pipeline system regardless of advantages and disadvantages for individual pipeline segments. The operators told us that automatic-shutoff valves, as opposed to remote-control valves, were chosen because they reduce the potential for human error when making decisions to close valves. Officials stated that the biggest concern of using automatic-shutoff valves is the potential for accidental closures, but they believe they have developed a procedure for managing the pressure sensing system that effectively adapts to pressure and flow change and minimizes or eliminates these types of closures. Monetary costs are not considered as part of the decision-making process. The operator stated that installing automatic-shutoff valve on an existing manual valve costs between $48,000 and $100,000.

Results to date: According to Kinder Morgan officials, this approach has resulted in the installation of automated valves at 683 out of 832 locations across the pipeline system. Officials plan to automate the remaining valves over the next several years.

Pipeline operator: Northwest Pipeline GP[4]

Product type: Natural gas

Number of pipeline miles: 3,900 (total); 170 (high-consequence areas)

Decision-making approach: The operator uses a decision tree to assess individual pipeline segments based on several criteria, including the location of the valve (e.g., high-consequence area), diameter of the pipe, and the amount of time it takes for an operator to respond upon notification of an incident. The operator will install an automated valve in any high-consequence, class 3, or class 4 areas[5] on large diameter pipe (i.e., above 12 inches) where personnel cannot reach and close the valve in under an hour. Monetary costs are considered as part of the decision-making process for the purposes of determining the most cost-effective way to ensure the operator can respond within one hour to incidents in high-consequence areas. The operator stated that installing an automated valve costs between $37,000 and $240,000. Automatic-shutoff valves are not installed in areas where an accidental closure could lead to customers losing service (i.e., in places where there is a single line feed servicing the entire area) or where pressure fluctuations may inadvertently activate the valve.

Results to date: According to Northwest Pipeline GP officials, this approach has resulted in decisions to install automated valves at 59 of the 730 locations assessed.

Pipeline operator: Williams Gas Pipeline-Transco[6]

Product type: Natural gas

Number of pipeline miles: 11,000 (total); 1,192 (high-consequence areas)

Description of decision-making method: The operator uses a decision tree to assess individual pipeline segments based on several criteria, including the location of the valve (e.g., high-consequence area), diameter of the pipe, and the amount of time it takes for an operator to respond upon notification of an incident. The operator will install an automated valve in any high-consequence, class 3, or class 4 areas on large diameter pipe (i.e., above 12 inches) where personnel cannot reach and close the valve in under an hour. Monetary costs are considered as part of the decision-making process for the purposes of determining the most cost-effective way to ensure the operator can respond within one hour to incidents in high-consequence areas. The operator stated that installing an automated valve costs between $75,000 and $500,000. Automatic-shutoff valves are not installed in areas where an accidental closure could lead to customers losing service (i.e., in places where there is a single line feed servicing the entire area) or where pressure fluctuations may inadvertently activate the valve.

Results to date: According to Williams Gas Pipeline-Transco officials, this approach has resulted in decisions to install automated valves at 56 of the 2,461 locations assessed.

APPENDIX III: AUTOMATED VALVE COSTS

The eight operators we spoke with provided a range of cost estimates for installing automated valves—from as low as $35,000 to as high as $500,000 depending on the location and size of the pipeline, and the type of equipment being installed, among other things.[1]

While both hazardous liquid and natural gas transmission pipeline operators estimated a similar cost range from about $35,000 to $500,000, hazardous liquid pipeline operators tended to estimate higher costs. Specifically, two of the three operators that exclusively transport hazardous liquids estimated that the minimum costs of installing an automated valve was $100,000 or higher and the maximum was $500,000.

In contrast, pipeline operators that exclusively transport natural gas all estimated that the minimum cost was $75,000 or lower and three of the four operators estimated that maximum costs would be $240,000 or lower.

We also spoke with five equipment vendors and six contractors that install valves to gather additional perspective on the cost of purchasing and installing automated valve equipment.[2]

According to estimates provided by these businesses, the combined equipment and labor costs range between $40,000 and $380,000. Specifically, equipment costs range from $10,000 to $75,000 while labor costs range from $30,000 to $315,000. (See table 3.)

Vendors stated that the cost of installing an automated valve depends primarily on the functionality of the equipment (for example, additional controls would increase the cost), while contractors stated that these costs depend on the diameter and location of the pipeline. Vendors and contractors had varying opinions on whether the costs were greater to install an automated valve on hazardous liquid or natural gas pipeline.

Table 3. Range of Equipment and Labor Costs, According to Pipeline Vendors and Contractors

Equipment vendor	Range of equipment costs	Contractor	Range of labor costs
#1	$10,000 – $75,000	#1	$30,000 – $150,000
#2	$15,000 – $30,000	#2	$35,000 – $175,000
#3	$15,500 – $43,000	#3	$40,000 – $150,000
#4	$30,500 – $43,500	#4	$100,000 – $200,000
#5	$32,000 – $46,000	#5	$120,000 – $315,000
Average	$20,600 – $47,500	Average	$65,000 – $198,000

Source: GAO presentation of vendor and contractor information.

End Notes

[1] In its regulations, PHMSA refers to the release of natural gas from a pipeline as an "incident" (49 C.F.R. § 191.3) and a spill from a hazardous liquid pipeline as an "accident." (49 C.F.R. §195.50). For simplicity, this report will refer to both as "incidents."

[2] "High-consequence areas" are defined differently for hazardous liquid and natural gas. For natural gas, such areas typically include highly populated or frequented areas, such as parks. For hazardous liquid, high-consequence areas include highly populated areas, other populated areas, navigable waterways, and areas unusually sensitive to environmental damage.

[3] For the purposes of this report we use the term "install an automated valve" to refer to any actions that allow the operator to remotely or automatically close a valve. Such actions do not necessarily mean an operator is installing a completely new valve. For example, operators may install an actuator and communications at an existing valve location.

[4] See NTSB, *Pipeline Accident Report: Pacific Gas and Electric Company Natural Gas Transmission Pipeline Rupture and Fire, San Bruno, California*, September 9, 2010, NTSB/PAR-11/01 (Washington, D.C: Aug. 30, 2011). According to NTSB, PHMSA is in the process of responding to this recommendation. Specifically, in August 2011, PHMSA began a rulemaking process that could address the extent to which operators will be required to install automated valves. 76 Fed. Reg. 53086 (Aug. 25, 2011).

[5] For the purposes of this report, we use the term "transmission pipeline" to refer to both onshore hazardous liquid and natural gas pipelines carrying product over long distances to users.

[6] The Act also directed the Secretary of Transportation to consider additional regulations requiring the use of automated valves where economically, technically, and operationally feasible on new transmission facilities. Pub. L. No. 112-90, § 4, 125 Stat. 1904, 1906 (2012). In response, PHMSA contracted with Oak Ridge National Laboratory to draft a study, which found that automated valves were feasible under certain conditions. Oak Ridge National Laboratory, *Studies for the Requirements of Automatic and Remotely Controlled Shutoff Valves on Hazardous Liquids and Natural Gas Pipelines with Respect to Public and Environmental Safety*, ORNL/TM-2012/411 (Oct. 31, 2012).

[7] According to 2010 PHMSA data, the eight operators we selected represented 19 percent of hazardous liquid and 10 percent of natural gas miles in these areas. There were 682

hazardous liquid and natural gas transmission pipeline operators with 98,013 pipeline miles in high-consequence areas.

[8] Hazardous liquid products include petroleum (crude oil, condensate, natural gasoline, natural gas liquids, and liquefied petroleum gas); petroleum products (flammable, toxic, or corrosive products obtained from distilling and processing of crude oil, unfinished oils, natural gas liquids, blend stocks, and other miscellaneous hydrocarbon compounds); and anhydrous ammonia.

[9] PHMSA does not regulate all pipelines. For example, many gathering pipelines have not been subject to PHMSA regulations because they are generally located away from population centers and operate at low pressures.

[10] PHMSA may conduct an incident investigation in instances when an NTSB investigation is also under way. In such cases, PHMSA does not determine the cause of the incident; rather its review is to determine regulatory compliance.

[11] PHMSA established requirements (49 C.F.R. § 195.452) for integrity management for hazardous liquid pipeline operators with 500 or more miles of pipelines in December 2000 (65 Fed. Reg. 75378, (Dec. 1, 2000)) and for operators with less than 500 miles in January 2002 (67 Fed. Reg. 2136, (Jan.16, 2002)). In 2003, PHMSA issued integrity management regulations for all operators of gas transmission pipelines (68 Fed. Reg. 69778, (Dec.15, 2003)).

[12] *Warning letters* are issued for lower risk probable violations and program deficiencies. Through such letters PHMSA notifies the operator of the alleged violations and directs it to correct them or be subject to further enforcement action. *Notices of probable violation* allege specific regulatory violations and, where applicable, propose corrective action in a compliance order and/or civil penalties. The operator has a right to respond and request an administrative hearing. *Notices of amendment* allege that an operator's plans and procedures are inadequate and require that they be amended. The operator has a right to respond and request an administrative hearing. *Notices of proposed safety order* notify an operator that a particular pipeline facility has a condition or conditions that pose a pipeline integrity risk to public safety, property, or the environment. These notices propose measures the operator must take to address the identified risk, including inspection, testing, and repair. *Corrective action orders* are issued to operators with a pipeline that represents a serious hazard to life, property, or the environment. The order identifies actions that must be taken by the operator to assure safe operation, including the shutdown of a pipeline or operation at reduced pressure, physical inspection or testing of the pipeline, and repair or replacement of defective pipeline segments, among other actions.

[13] The risks and consequences posed by gas and hazardous liquids incidents also differ. Natural gas tends to ignite more easily, resulting in more explosions. Hazardous liquids ignite less easily, but can spill and pollute the environment.

[14] The National Response Center is the sole federal point of contact for reporting oil and chemical spills.

[15] Hazardous liquid regulations refer to emergency flow restriction devices, which include remote-control valves and "check" valves that automatically prevent product from flowing in a specific direction. See 49 C.F.R. § 195.452(i)(4). For the purposes of this report we describe all of these valves as automated valves.

[16] PHMSA does not collect data on the number of manual and automated valves. The Interstate Natural Gas Association of America—the primary industry group for natural gas transmission pipelines—has collected valve equipment information for almost half of the 300,000 miles of natural gas transmission pipeline in the United States and reports that of

the 29,827 valves reported 5,013, or 17 percent, are automated. In highly populated and frequented locations 1,972, or 23 percent, of the 8,693 total valves, were automated.

[17] 49 C.F.R. §§ 192.179, 195.260.

[18] Automated valves are one of several measures that operators can take to prevent and mitigate the consequences of a pipeline incident. Other measures include additional leak detection and damage prevention activities.

[19] Hazardous liquid pipeline operators are required to have a leak detection system on their pipeline. Natural gas pipeline operators may choose to install a leak detection system, although they are required to periodically survey their pipeline to identify leaks.

[20] PHMSA requires pipeline operators to develop and follow written control room management procedures that define the roles and responsibilities of control room personnel in normal, abnormal, and emergency operating situations. This requirement allows individual operators to define the specific responsibilities for control room management by considering the characteristics of the operator's pipeline and its methods of safely managing pipeline operation.

[21] PHMSA requires pipeline operators to establish and maintain communications with fire, police, and other appropriate public officials to learn the responsibility and resources of each government organization that may respond to a natural gas or hazardous liquid pipeline emergency and acquaint the officials with the operator's ability in responding to an emergency. Operators must also plan and coordinate their responses to emergency incidents with these officials.

[22] A right of way is a strip of land, usually between 25 to 150 feet wide, containing one or more pipelines.

[23] We consider performance-based regulations those that focus on desired, measurable outcomes, rather than prescriptive processes, techniques, or procedures.

[24] 76 Fed. Reg. 3821, § 1 (a), (b)(4) (Jan. 21, 2011).

[25] GAO, *Statewide Transportation Planning: Opportunities Exist to Transition to Performance-Based Planning and Federal Oversight*, GAO-11-77 (Washington, D.C.: December 2010).

[26] In addition, NTSB has recommended that the Department of Transportation conduct an audit to assess the effectiveness of PHMSA's oversight of performance-based safety programs. See NTSB, *Pipeline Accident Report: Pacific Gas and Electric Company Natural Gas Transmission Pipeline Rupture and Fire, San Bruno, California*, September 9, 2010, NTSB/PAR-11/01 (Washington, D.C: Aug. 30, 2011). In response to the NTSB recommendation, the Department of Transportation is currently conducting an audit, which it expects to issue in early 2013, that will evaluate the effectiveness of PHMSA's inspection and oversight of pipeline operators' integrity management programs, including expanding the use of meaningful metrics and setting goals for pipeline operators and tracking performance against those goals.

[27] Emergencies include natural gas detected inside or near a building, accidental release of hazardous liquid or carbon dioxide from a pipeline facility, fire or explosion occurring near or directly involving a pipeline facility, operational failure causing a hazardous condition, or natural disaster affecting pipeline facilities.

[28] According to these standards, gas distribution companies have three response time targets—30 minutes, 45 minutes, and 60 minutes—which companies must meet between 76 to 97 percent of the time, depending on the response time target and the operator's working hours when the call is received (i.e., normal business hours, after hours, and weekends/holidays). For example, gas distribution companies are expected to achieve a 30-minute response 76 percent of the time during weekends and holidays, but 82 percent of the time during normal

business hours. These response time standards also apply to other events, such as odor complaints or reports of damage to the pipeline.

[29] According to officials from the Interstate Natural Gas Association of America, which represents natural gas transmission pipeline operators, members will conduct a risk analysis on a case-by-case basis to determine the appropriate maximum incident response time for small diameter (i.e., 12 inches or less) natural gas pipelines in highly populated areas. Prior to the 1-hour goal for large diameter pipelines, members did not have any incident response time commitment.

[30] We have reported on the need for comprehensive and reliable data to implement a performance-based approach. See GAO, *Surface Transportation: Restructured Federal Approach Needed for More Focused, Performance-Based, and Sustainable Programs,* GAO-08-400 (Washington, D.C.: Mar. 6, 2008).

[31] The DOT Inspector General has also reported on significant data problems with PHMSA hazardous liquid pipeline operator data. In its 2012 audit, the DOT Inspector General identified shortcomings in PHMSA data management and quality that limit the usefulness of incident and annual report data. For example, according to the audit, PHMSA lacks a method to detect duplicate reporting of annual shipment volumes from one year to the next. PHMSA reported that it created a data quality assurance plan in 2010 and has implemented many improvements in its pipeline safety data systems, but still faces significant staffing and funding needs to make further improvements. See Office of Inspector General, U.S. Department of Transportation, *Hazardous Liquid Pipeline Operators' Integrity Management Programs Need More Rigorous PHMSA Oversight*, AV-2012-140 (Washington, D.C.: June, 2012).

[32] PHMSA guidance states that when reporting on when an operator identified the incident, operators should enter the date and time when the operator became aware or first identified when the incident had occurred, and not when the operator determined that the incident met the reporting criteria.

[33] NTSB, *Pipeline Accident Report: Pacific Gas and Electric Company Natural Gas Transmission Pipeline Rupture and Fire, San Bruno, California*, September 9, 2010, NTSB/PAR-11/01 (Washington, D.C: Aug. 30, 2011).

[34] See below and appendix III for a discussion on the costs of automated valves.

[35] See appendix II for more details on the methods used by each operator to determine whether or not to install automated valves.

[36] For a discussion of the costs of installing automated valves see appendix III.

[37] Federal regulations require hazardous liquid and natural gas pipeline operators to consider measures to prevent and mitigate the consequences of a pipeline failure that could affect a high-consequence area, including installing automated valves on individual pipeline segments if the operator determines such a valve would add protection and enhance public safety. As part of this determination, operators must consider certain factors at a minimum. These factors relate to descriptive characteristics of individual pipeline segments—such as pipeline profile and operating pressure—and consideration of the possible safety and environmental outcomes. See 49 C.F.R. §§ 192.935, 195.452(i).

[38] GAO, *Pipeline Safety: Collecting Data and Sharing Information on Federally Unregulated Gathering Pipelines Could Help Enhance Safety*, GAO-12-388 (Washington, D.C.: Mar. 22, 2012) and GAO, *Rail Transit: FTA Programs Are Helping Address Transit Agencies' Safety Challenges, but Improved Performance Goals and Measures Could Better Focus Efforts*, GAO-11-199 (Washington, D.C.: Jan. 31, 2011).

[39] See http://primis.phmsa.dot.gov/comm/.

End Notes for Appendix I

[1] According to 2010 PHMSA data, the eight operators we selected represented 19 percent of hazardous liquid and 10 percent of natural gas miles in high-consequence areas. There were 682 hazardous liquid and natural gas transmission pipeline operators with 98,013 pipeline miles in high-consequence areas.

End Notes for Appendix II

[1] Requirements for integrity management require hazardous liquid pipeline operators to determine whether an incident on their pipeline could affect a high-consequence area, while natural gas pipeline operators are required to determine whether their pipeline is in a high-consequence area. Regulations also define how to identify high-consequence areas for these two types of operators.

[2] Pipeline segments are discrete sections of the pipeline system separated by valves that can stop the flow of product. The distance between valves is dictated in federal regulations. C.F.R. §§ 192.179, 195.260.

[3] According to the operator, several automated valves have been installed independent from this decision-making approach.

[4] Williams Gas Pipeline is the parent company of two operators we visited: Northwest Pipeline GP and Williams Gas Pipeline-Transco. Northwest Pipeline GP is also referred to as WGP West. Williams Gas Pipeline-Transco is one of three pipeline systems that make up WGP East, which also includes Gulfstream Pipeline and Cardinal Pipeline.

[5] The Pipeline and Hazardous Materials Administration regulates natural gas pipelines based on class locations. Class 3 includes any location with more than 46 buildings intended for human occupancy within 220 yards of a pipeline, or an area where the pipeline lies within 100 yards of either a building or a small, well-defined outside area (such as a playground) that is occupied by 20 or more persons at least 5 days a week for 10 weeks in any 12-month period. Class 4 includes any location where unit buildings with four or more stories above ground are prevalent. See 49 CFR 192.5.

[6] Williams Gas Pipeline is the parent company of two operators we visited: Northwest Pipeline GP and Williams Gas Pipeline-Transco. Northwest Pipeline GP is also referred to as WGP West. Williams Gas Pipeline Transco is one of three pipeline systems that make up WGP East, which also includes Gulfstream Pipeline and Cardinal Pipeline.

End Notes for Appendix III

[1] Several hazardous liquid operators indicated that the cost of installing an automated valve could be as high as $1 million under specific circumstances, but stated that these high costs are unusual.

[2] One pipeline contractor we spoke with had not yet installed an automated valve and did not did not provide cost estimates for doing so. Instead the contractor stated that the labor costs of installing a manual valve on an existing pipeline would be between $60,000 and $80,000.

In: Pipeline Safety
Editor: Maeve B. Reagan

ISBN: 978-1-62618-337-7

Chapter 2

STATEMENT OF SENATOR JOHN D. (JAY) ROCKEFELLER IV, CHAIRMAN, U.S. SENATE COMMITTEE ON COMMERCE, SCIENCE, AND TRANSPORTATION. HEARING ON "PIPELINE SAFETY: AN ON-THE-GROUND LOOK AT SAFEGUARDING THE PUBLIC"*

They crisscross underneath our cities and country sides, yet most of the time we are not even aware they are there. They deliver critical fuel that powers our homes, factories, and offices; and also transport the oil and gas that keep our cars, trucks, and planes operating. They are the critical conduit between the shale gas development boom in our region and the rest of the country. Most days, the network of pipelines operates across the country without a hitch. Compared to other forms of transportation, pipelines are a relatively safe, clean and efficient way of transporting the goods they carry. Unfortunately, this is not always the case.

Everyone in this room knows all too well what can happen when something goes wrong. Last month's incident in Sissonville was a startling reminder of the destruction that can occur when a pipeline ruptures. Houses were destroyed and portions of the nearby interstate were literally

* This is an edited, reformatted and augmented version of a statement presented January 28, 2013 before the Senate Committee on Commerce, Science, and Transportation.

disintegrated by the overwhelming heat of flames from the ruptured pipe. We can only thank our lucky stars that there were no lives lost or serious injuries. As we have seen with other accidents in the last few years, we are not always so lucky. After the explosion in Sissonville, we must sustain our focus on making sure the pipeline industry and all industries operate as safely as possible. While we do not yet know the exact cause of the Sissonville incident, today's hearing provides the opportunity to examine where we stand in regard to the safety of our nation's pipeline system.

When I took over as Chairman of the Commerce Committee, I made consumer protection and public safety my key priorities. Accordingly, the Committee has been very active on the safety front and these efforts have resulted in safety improvements across several industries, from aviation to trucking to automobiles. As for pipeline safety, the Committee has held multiple hearings and successfully worked with our colleagues in the House to pass a pipeline safety bill into law just over one year ago. This law – The Pipeline Safety, Regulatory Certainty, and Job Creation Act of 2011 – was largely based on legislation we passed out of the Commerce Committee and through the Senate. This legislation included a number of new requirements that will move the ball forward in pipeline safety.

For example, we laid the foundation to require the use of remote controlled and automatic shut-off valves on new pipelines – something I know we'll discuss today. We removed exemptions from requirements to call and get underground lines marked before you excavate. Year after year, excavation damage is the leading cause of pipeline accidents. Removing exemptions from who has to "call before you dig" will reduce this problem. We required operators to verify records and re-establish lines' maximum operating pressure. Lack of records about older pipelines is a real problem and contributed to a catastrophic pipeline explosion in California that killed several people. We required that critical pipeline location and inspection information be provided to the public to build greater awareness of the lines that exist in and around our communities. Finally, we increased penalties on operators who shirk the safety regulations. This will help deter bad actors from avoiding their safety obligations.

While I was pleased with the progress we made in last year's law, I pushed for stronger requirements to move pipeline safety even further ahead. The Senate-passed bill, while not perfect, included a number of more stringent requirements for operators, but House negotiators demanded watered-down provisions for an agreement to move forward. On the bright side, I'm confident that we will see strong, but fair and sensible, safety regulations out of our legislation. That's one of the things I want to discuss today with our federal panel of witnesses.

It was a tough fight to get pipeline safety legislation signed into law. However, it's important that we continue to provide rigorous oversight of the industry to determine whether serious gaps still exist in our safety requirements. Today is a perfect opportunity to take stock in where we are and consider what steps might be necessary moving forward.

As we consider what steps are necessary moving forward, we must remember these are crucial decisions and policies that have real impact on people's lives.

In: Pipeline Safety
Editor: Maeve B. Reagan

ISBN: 978-1-62618-337-7

Chapter 3

STATEMENT OF CYNTHIA L. QUARTERMAN, ADMINISTRATOR, PIPELINE AND HAZARDOUS MATERIALS SAFETY ADMINISTRATION. HEARING ON "PIPELINE SAFETY: AN ON-THE-GROUND LOOK AT SAFEGUARDING THE PUBLIC"*

Chairman Rockefeller and members of the Committee, thank you for the opportunity to appear today to discuss the progress the Pipeline and Hazardous Materials Safety Administration (PHMSA) has made to implement the mandates of the Pipeline Safety, Regulatory Certainty, and Job Creation Act of 2011 (Pipeline Safety Act).

Thank you for your leadership in helping to secure passage of the Pipeline Safety Act and for your efforts to advance pipeline safety. The Act has given us important tools and authority that we need to help us achieve our mission. While pipeline safety is improving, high-profile incidents like the one that occurred at Sissonville underscore how important it is to be ever-vigilant in preventing pipeline failures.

Safety is the top priority for Secretary of Transportation Ray LaHood and myself, and everyone at PHMSA is working hard to protect the American people and environment from the risks that are inherent in the transportation of hazardous materials by pipeline. PHMSA works to achieve its safety mission

---

* This is an edited, reformatted and augmented version of a statement presented Jan 28, 2013 before the Senate Committee on Commerce, Science and Transportation.

through prevention, rigorous enforcement, strong partnerships, and continuing education.

This testimony will focus on several issues such as to the implementation of the Pipeline Safety Act mandates; our response to the Sissionville, WV pipeline incident and the Government Accountability Office (GAO) study on the ability of transmission pipeline facility operators to respond to a hazardous liquid or gas release.

First, I will give an overview of PHMSA's pipeline safety program, including the role that the States take in ensuring the safety of pipelines. Second, I will provide an overview of the mandates we have completed and the efforts we have taken to improve pipeline safety. Third, I will discuss how, incidents like the one at Sissonville show us that we have a long way to go to succeed in our mission and that there is still a lot of work to be done in preventing pipeline incidents. Finally, I will reiterate the importance of a robust pipeline safety program, and the importance of reviewing the findings of the GAO study especially with regard to the Nation's changing and growing infrastructure needs.

I. Overview of PHMSA Pipeline Safety Program

There are 2.6 million miles of pipelines that crisscross our Nation; those pipelines offer the safest and most cost-efficient way to transport hazardous materials. To ensure that this vast network is operating safely and reliably and that communities and families are protected, PHMSA works together with a variety of partners, including other Federal agencies, State and local officials, emergency responders, environmental groups, and the public.

Federal oversight agencies like the National Transportation Safety Board (NTSB), the Office of Inspector General (OIG), and the Government Accountability Office (GAO) also have a vested interest in the safe and reliable operation of the Nation's pipeline infrastructure. For years, we have worked aggressively to respond to their recommendations. In addition to the mandates of the Act, we are currently working on 26 open NTSB recommendations, 9 recommendations from the OIG, and 4 recommendations from the GAO. Some of these recommendations are similar to the requirements of the Pipeline Safety Act, which suggests that there is a shared understanding of some of the challenges for the Nation's pipeline system.

We have taken each and every mandate and recommendation that has been issued to us very seriously, and we have many completed and ongoing initiatives to provide protection to the American people and environment.

Overall, the pipeline safety record is good. PHMSA's regulatory oversight program has led to many successes. Despite the fact that the traditional measures of risk—population, energy consumption, pipeline ton-miles—have steadily increased over the past two decades, the risk of pipeline incidents with death or major injury have decreased by about 10 percent every 3 years. The risks of hazardous liquid pipeline spills that have environmental consequences have decreased by an average of 5 percent per year. Nonetheless, there is more work to be done.

In 2012, the number of pipeline-related fatalities was at a level not seen since 2008, and the number of pipeline-related injuries was at the lowest level since 2007. Furthermore, 2012 had the fewest total pipeline incidents in a decade. However, PHMSA, as an organization, cannot accept death or injury as an inevitable consequence of transporting hazardous materials. We are working continuously to find new ways to reduce risk to operators and the public, and we aim to sustain and improve upon these long-term trends.

II. Implementation of the Pipeline Safety Act

On January 3, 2012, President Obama signed the Pipeline Safety, Regulatory Certainty, and Job Creation Act of 2011. The Act is designed to examine and improve the state of pipeline safety regulations and authorizes funding, through fiscal year 2015, for provisions of the pipeline statute in the U.S. Code related to gas and hazardous liquids. Ultimately, the Act gives enhanced safety authority to PHMSA and will improve pipeline transportation, by strengthening the enforcement capabilities of current laws.

The leadership of Chairman Rockefeller and this committee, as well as the bipartisan effort that led to the creation and passage of the Pipeline Safety Act shows there is a common agreement about the importance of a safe and reliable pipeline system for the welfare of the Nation. PHMSA takes this responsibility very seriously. As the committee is aware, we have struggled to hire pipeline inspectors over the last several years, but by the end of FY 2012, we achieved and successfully filled our targeted 135 pipeline inspector billets. We now look forward to working with this committee to continue to strengthen our pipeline inspector program and further implement PHMSA's Pipeline Safety Reform effort.

PHMSA not only completed all of the mandates that were due by January 3, 2013, it also completed additional mandates and performed more work than required. PHMSA has already successfully completed 16 of the 42 requirements in the Pipeline Safety Act. PHMSA has reported on cover over buried pipelines at river crossings, leak detection, remote controlled and automatic shut-off valve (RCV/ASV) use, increasing civil penalties authority, improved the quantity, quality, and transparency of our data, and inventoried the status of cast iron pipeline infrastructure. Information gathered from these reports will be used to inform us as we determine how best to move forward with updated requirements to address these topics.

The following is a brief description of PHMSA's work the Pipeline Safety Act requirements:

Section 2—Civil Penalties

The Act authorized PHMSA to increase the maximum civil penalty for pipeline safety violations from $100,000 to $200,000 per violation per day. In addition, the agency will be able to collect a maximum of $2,000,000 for a related series of violations, up from $1,000,000.

PHMSA is currently addressing this activity through a rulemaking to update Part 190 of the Code of Federal Regulations. A Notice of Proposed Rulemaking (NPRM) entitled "Administrative Procedures; Updates and Technical Corrections" was published on August 13, 2012.

Section 3—Pipeline Damage Prevention

The Act required PHMSA to incorporate new standards for state one-call programs into the State Damage Prevention (SDP) grant program criteria, including no state and local exemptions.

Some state excavation damage prevention laws include exemptions from one-call system participation that detract from the goals of the system. The following are examples of two typical types of exemption:

> Facility Owners—some state laws exempt owners of specific types of underground facilities (e.g., municipalities, State departments of transportation, and small water and sewer companies from participation in the one-call system). Excavators—some excavators (e.g., homeowners

and State departments of transportation) are exempted from calling for underground facilities to be located and marked before they begin digging. PHMSA has discussed these exemptions with the National Association of Pipeline Safety Representatives (NAPSR) and One Call Systems International (OCSI). A public meeting regarding these issues is scheduled for March 2013. These new requirements were included in the SDP grant program criteria.

The Act also requires for PHMSA to conduct a study on the impact of excavation damage on pipeline safety, including exemptions, frequency, severity, and type of damage, and report these results to Congress.

PHMSA met with the United States Infrastructure Corporation (USIC) to discuss performing a data analysis regarding damage prevention. As mentioned above, PHMSA is planning a public meeting in March 2013 to discuss damage prevention issues with industry stakeholders.

PHMSA is considering using data from the Common Ground Alliance's (CGA's) Damage Information Reporting Tool (DIRT) to help with this study it will reach out to states to discuss the use of this data in the analysis.

Section 4—Automatic and Remote-Controlled Shut-off Valve Use

The Act requires PHMSA to issue regulations requiring the use of automatic or remote-control shut-off valves on transmission pipelines constructed or entirely replaced after the date of the rule, if appropriate.

PHMSA began to collect information on the use of automatic shut-off valves (ASVs) and remote-controlled shut-off valves (RCVs) on hazardous liquid and gas transmission pipelines prior to the enactment of the Pipeline Safety Act, through issuance of two Advanced Notice of Proposed Rulemakings (ANPRM) entitled "Safety of On-Shore Hazardous Liquid Pipelines" and "Safety of Gas Transmission Pipelines". For hazardous liquid transmission pipelines, an ANPRM issued on October 18, 2010, requested public comments on the use of RCVs. For gas transmission pipelines, an ANPRM issued on October 25, 2011, requested public comments on requiring the use of ASV and RCV installation.

To gather sufficient input on ASV/RCV feasibility, PHMSA sponsored a public workshop on March 28, 2012 with the National Association of Pipeline Safety Representatives, entitled "Understanding the Application of Automatic Control & Remote Control Valves." PHMSA then commissioned an independent study on the feasibility and effectiveness of ASVs and RCVs on hazardous liquid and natural gas transmission pipelines. Public comments and workshop input were used to develop the commissioned study entitled, "Studies for the Requirements of Automatic and Remotely Controlled Shutoff Valves on Hazardous Liquid and Natural Gas Pipelines with Respect to Public and Environmental Safety" (ASV-RCV study), including the original scope of work.

The ASV-RCV study performed by the Oak Ridge National Laboratory, while not mandated by the Act, will help to determine the effectiveness of block valve closure swiftness in mitigating the consequences of natural gas and hazardous liquid pipeline releases on the safety of the public and the environment. Additionally, a related NTSB recommendation, NTSB P-11-11, was incorporated into the parameters of the study. The recommendation suggested ASVs and RCVs be required in high-consequence areas (HCAs). A public web-based seminar (webinar) and public comment period was also held for input on the draft study. The ASV-RCV study addressed the submitted comments and incorporated substantive technical recommendations. The ASV-RCV study, which is 344 pages, was transmitted to Congress on December 27, 2012.

The information from this study will assist in providing additional guidance for potential rulemaking. PHMSA also anticipates progressing with a rulemaking related to ASV and RCV installation and use on hazardous liquid and gas transmission pipelines in 2013.

In addition, PHMSA is soliciting a research project specific to technology used in ASVs that will provide important insight on their ability to provide reliability and flow assurance to pipelines. Automatic shut-off valves are often recommended to minimize valve shut-off times after a leak is detected. However, they may lead to unintended valve closures because of an inaccurate leak determination. The project aims to study and identify technologies and systems to minimize inaccurate leak alarms and unintended valve closures on ASV systems.

Section 5—Integrity Management

The Act required PHMSA to conduct an evaluation on whether integrity management programs (IMPs) should be expanded beyond high-consequence areas (HCAs) and whether gas IMPs should replace the class location system.

This section also asks, PHMSA to consider issuing regulations expanding IMP requirements and/or replacing class locations.

As mentioned above, PHMSA initiated an ANPRM, entitled "Safety of On-Shore Hazardous Liquid Pipelines" and "Safety of Gas Transmission Pipelines" for both gas and liquid pipeline safety that addresses these issues. PHMSA is also holding an integrity management program (IMP) 2.0 workshop in 2013.

This section of the statute also suggests that PHMSA may extend a gas pipeline operator's 7-year reassessment interval by 6 months if the operator submits written notice with sufficient justification of the need for an extension, and that PHMSA should publish guidance on what constitutes sufficient justification. PHMSA is currently considering this issue in the context of a gas transmission NPRM, which is a follow on from the ANPRM entitled "Safety of Gas Transmission Pipelines" mentioned above. PHMSA anticipates this NPRM to be published by August 2013.

Section 6—Public Education and Awareness

There were several mandates in this section of the Act. One mandate requires that PHMSA maintain a map of all gas HCAs as a part of the National Pipeline Mapping System (NPMS). PHMSA has already begun implementing this with the information we have currently available, and we are continuing to work on expanding the information available. PHMSA was also requested to update the NPMS map biennially.

In addition, PHMSA was required to implement a program for promoting greater awareness of the NPMS to state and local emergency responders and other parties. To address this issue, PHMSA hosted a meeting of Public Safety and Emergency Response officials to discuss pipeline emergency preparedness and response on December 9, 2011. Additionally, PHMSA made contact with various emergency responder groups through its Emergency Responder (ER) Outreach program and the Community Assistance and Technical Services (CATS) program. PHMSA has also begun publishing articles regarding its public resources, including the NPMS, in ER publications. A brochure, designed for widespread distribution in the ER community, was also created that described available resources.

PHMSA was also required to issue guidance to operators to provide system-specific information about their pipelines to emergency responders after consulting with those responders. This mandate fell closely in line with an NTSB recommendation (P-11-8), which recommended pipe diameter, operating pressure, product transported, and potential impact radius, among other information, is shared.

PHMSA, in partnership with the Pipeline Emergency Response Working Group (PERWG), met with emergency responders at a pipeline emergency response focus group during the HOTZONE conference in Houston on October 19, 2012.

The PERWG had its follow up meeting last week. On October 11, 2012, PHMSA published (Advisory Bulletin ADB-12-09) about Communication During Emergency Situations that reminds operators of gas, hazardous liquid, and liquefied natural gas pipeline facilities that operators should immediately and directly notify the Public Safety Access Point that serves the communities and jurisdictions in which those pipelines are located when there are indications of a pipeline facility emergency. We also met with the Associate of Public Communication Offices to discuss how to increase awareness and develop training for 911 center personnel.

Additionally, PHMSA is funding a Transportation Research Board study that will produce a guide for communication between pipeline operators and emergency responders.

PHMSA recognizes and agrees that the emergency response to an incident or a leak is critical. In addition to strengthening the capabilities of local emergency responders with increased coordination, targeted planning, and training grants. PHMSA has also worked to increase the visibility of prevention and response efforts to better prepare the public.

The final mandate from this section required PHMSA to maintain the most recent oil facility response plans (FRPs), which are currently collected from operators and provide copies of those FRPs to any requester through the FOIA process. The copies can exclude sensitive information. PHMSA has implemented this mandate and continues to improve the FRP program.

Section 7—Cast Iron Gas Pipelines

The Act required PHMSA to follow-up on the industry's progress in replacing cast iron gas pipelines. PHMSA has collected updates and has published the responses on its website which can be found at http://opsweb.phmsa.dot.gov/pipelineforum/. This inventory was developed and posted before the December 31, 2012 due date.

Section 8—Leak Detection

The Act requires PHMSA to submit a report to Congress on leak detection systems used by operators of hazardous liquid pipeline facilities and transportation related flow lines. The Act requires the following be included in the report:

- an analysis of the technical limitations of current leak detection systems, including the ability of the systems to detect ruptures and small leaks that are ongoing or intermittent, and what can be done to foster development of better technologies; and
- an analysis of the practicability of establishing technically, operationally, and economically feasible standards for the capability of such systems to detect leaks, and the safety benefits and adverse consequences of requiring operators to use leak detection systems.

PHMSA began working on leak detection for a number of years before the Act. As mentioned above, on October 18, 2010, an ANPRM for the Safety of On-Shore Hazardous Liquid Pipelines was published. Among the issues discussed in the ANPRM was whether to establish and/or adopt standards and procedures for minimum leak detection requirements for all pipelines.

In addition, PHMSA sponsored a public workshop in March 2012 with the National Association of Pipeline Safety Representatives entitled "Improving Pipeline Leak Detection System Effectiveness." It also held a Pipeline Research and Development (R&D) Forum in July 2012 that included a working group discussion focused specifically on leak detection and mitigation. As a result, PHMSA has issued a research announcement and solicitation for proposals for research and development on a number of topics, including leak detection. As part of its research and development activities, PHMSA has been active in studying and improving other leak detection technologies, including automated monitoring systems, sensors for small leak detection, aerial surveillance, satellite imaging, and improvements in the cost and effectiveness of current leak detection systems.

As with valves, PHMSA also commissioned an independent study on leak detection.

In conjunction with satisfying the requirements of the Act, PHMSA is also addressing a leak detection related recommendation for natural gas transmission and distribution pipelines from the NTSB (NTSB recommendation P-11-10, which involves Supervisory Control and Data

Acquisition (SCADA) enhancements to Identify and Locate Leaks). PHMSA's leak detection work included systems used in gas transmission and distribution pipelines as well as hazardous liquid pipelines. While the different types of pipeline systems have various and distinct characteristics and considerations for leak detection, PHMSA brought all pipeline industry stakeholders together to more efficiently communicate the issues affecting the respective sectors and to share lessons learned.

The review of leak detection systems was not limited to the technology but also extended to pipeline facilities and infrastructure. Effective leak detection relies heavily on how well any technology is implemented through people, procedures, and the environment in which it is installed and operated.

The leak detection study performed was based on input received through the workshops and a public comment period for the original scope of work. A public web-based seminar (webinar) and public comment period was also held for input on the draft report of the study. Additionally, some operators were interviewed as part of the work. The final leak detection study, which is almost 300 pages, has been posted electronically for review and has been transmitted to Congress.

PHMSA will use all of the input gathered from the above initiatives as well as other data when considering any future rulemakings. A rulemaking is under consideration for this item.

PHMSA is also creating a Leak Detection webpage on the PHMSA website to provide background information about leak detection issues.

Section 9—Accident and Investigation Notification

PHMSA was required by the Act to revise regulations to require telephonic reporting of incidents or accidents not later than 1 hour following a "confirmed discovery" and to require revising the initial telephonic report after 48 hours if practicable.

An NPRM entitled "Miscellaneous Rule II" regarding these revisions is expected to be issued in late 2013.

The Act also requires PHMSA to review and revise, as necessary, procedures for operators and the National Response Center (NRC) to notify emergency responders, including local public safety answering points or 911 centers.

PHMSA is continuing to develop a means to address this issue.

Section 10—Transportation-Related Onshore Facility Response Plan Compliance: Administrative Enforcement and Civil Penalties

While there was no specific mandate with this item, the section did suggest that PHMSA should update Part 190 to be consistent with the new authority to enforce Part 194 regulations. A rulemaking entitled "Administrative Procedures; Updates and Technical Corrections" is under consideration for this item.

Section 11—Pipeline Infrastructure Data Collection

PHMSA is considering collecting other geospatial and technical data for the NPMS. Although there was no specific mandate for this action, as mentioned in Section 11 above, a rulemaking is under consideration for this item.

Section 12—Transportation-Related Oil Flow Lines

There is no mandate related to this section, but PHMSA is considering collecting geospatial and other data on transportation-related oil flow lines, as mentioned in Section 11 above, as defined in the Act.

Section 13—Cost Recovery for Design Reviews

PHMSA was required to prescribe a fee structure and procedures for assessment and collection in order to implement authority to recover design review costs for projects that cost over $2.5 billion or that involve "new technologies."

PHMSA is currently developing guidance on this issue.

This section also mandates that PHMSA issue guidance on the meaning of the term "new technologies."

This guidance was completed and was posted on the external PHMSA website prior to the January 3, 2013 deadline.

Section 15—Carbon Dioxide Pipelines

The Act requires that PHMSA issue regulations for transporting carbon dioxide by pipeline in a gaseous state. PHMSA is currently exploring rulemaking options with this item.

Section 16—Study of Transportation of Diluted Bitumen

PHMSA was required to review and report to Congress on whether current regulations are sufficient to regulate pipelines transporting diluted bitumen. A study has been contracted to perform this analysis to the National Academy of Sciences (NAS), which is meeting on the issue on January 31 and February 1, 2013, and it is on track for timely completion. Once the study is completed, a report to Congress will follow.

Section 17—Study of Nonpetroleum Hazardous Liquids Transported by Pipeline

This section allows PHMSA to analyze the extent to which pipelines transporting non-petroleum hazardous liquids, such as chlorine, are unregulated, and whether being unregulated presents risks to the public. The results of any analysis must be made available to Congress as directed by the Act. PHMSA is currently reviewing this issue.

Section 19—Maintenance of Effort

PHMSA was required to grant waivers of the maintenance of effort clause in FY12 and FY13 to States that demonstrate an inability to maintain funding to their pipeline safety program due to economic hardship. This action has been completed for FY12, and we are addressing this issue as it pertains to future years.

Section 20—Administrative Enforcement Process

This section requires PHMSA to issue regulations for enforcement hearings that require a presiding official, implement a separation of functions, prohibit ex parte communications and provide other due process provisions. This issue is currently being addressed in the Part 190 Rule referred to in Section 20 above.

The NPRM entitled "Administrative Procedures; Updates and Technical Corrections" was published on August 13, 2012.

Section 21—Gas and Hazardous Liquid Gathering Lines

The Act requires PHMSA to review and report to Congress on existing Federal and State regulations for all gathering lines, existing exemptions, and the application of existing regulations to lines not presently regulated. PHMSA has contracted Oak Ridge National to assist in the research of this issue and a report is under development.

PHMSA must also consider issuing regulations that would subject offshore liquid gathering lines to the same standards as other liquid gathering lines. PHMSA will determine whether these regulations are necessary based on the results of the research and report.

Section 22—Excess Flow Valves

The Act requires PHMSA to consider issuing regulations requiring the use of excess flow valves on new or entirely replaced distribution branch services, multi-family facilities, and small commercial facilities.

PHMSA issued an ANPRM entitled "Expanding the Use of Excess Flow Valves in Gas Distribution Systems to Applications Other Than Single-Family Residences " on November 25, 2011 and is currently analyzing public comments.

Section 23—Maximum Allowable Operating Pressure (MAOP)

PHMSA was required to issue an Advisory Bulletin regarding the existing requirements to verify records confirming MAOP in Classes 3 and 4 and in

HCAs. An Advisory Bulletin on "Verification of Records" was issued for this item on May 7, 2012. PHMSA was also required to issue regulations requiring operators to report by July 3, 2013, any pipelines without sufficient records to confirm MAOP. As part of meeting the mandate, PHMSA determined they had the authority under existing regulations to collect this additional data. Therefore, PHMSA revised its gas transmission annual reporting form to collect this information which we will receive for the first time on June 15, 2013. The information collected will be used to address the mandate in the Act. This section also required PHMSA to issue regulations that require operators to report any exceedance of MAOP within 5 days, and to ensure the safety of pipelines without records to confirm MAOP. PHMSA published an advisory bulletin in the Federal Register on December 21, 2012 on Reporting the Exceedances of Maximum Allowable Operating Pressure (ADB2012-11). A rulemaking is under consideration for this item.

PHMSA was also required to issue regulations requiring tests to confirm the material strength of previously untested gas transmission pipelines in HCAs. As part of meeting the mandate, PHMSA determined they had the authority under existing regulations to collect this additional data. PHMSA will use its revised gas transmission annual report to collect this relevant data by June 15, 2013. This information will be used to meet the mandate in the Act.

Section 24—Limitation of Incorporation of Documents by Reference

This section requires PHMSA, starting in one year, to stop incorporating by reference into its regulations or guidance materials any industry standard unless it is publicly available free of charge on the internet. PHMSA is continuing to work with organizations that develop standards in order to make Incorporation-By-Reference (IBR) material available for free on the Internet. We are pleased that many standards setting organizations have agreed and are assisting PHMSA in complying with this item.

Section 28—Cover Over Buried Pipelines

PHMSA was required to conduct a study and report to Congress on hazardous liquid pipeline accidents at water crossings to determine if depth of cover was a factor. This study was completed and was transmitted to Congress before the January 3, 2013, deadline. If the study shows depth of cover was a factor, PHMSA must review the sufficiency of existing depth of cover regulations and consider possible regulatory changes and/or legislative recommendations. The Administration is still determining whether legislative changes should be recommended.

Section 29—Seismicity

There was no specific mandate within this section, but it was suggested that PHMSA should issue regulations to be consistent with the requirement in statute that operators consider seismicity in identifying and evaluating all potential threats to each pipeline pursuant to Parts 192 and 195. PHMSA has conducted research on this issue, which is currently under review.

Section 30—Tribal Consultation for Pipeline Projects

The Act requires PHMSA to develop and implement a protocol for consulting with Indian tribes to provide technical assistance for the regulation of pipelines that are under the jurisdiction of Indian tribes.

This protocol was posted on the PHMSA website prior to the January 3, 2013, deadline.

Section 31—Pipeline Inspection and Enforcement Needs

PHMSA was required to report to Congress on the total number of full-time equivalents (FTEs) for pipeline inspection and enforcement, the number of such FTEs that are not presently filled and the reasons they are not filled, the actions being taken to fill the FTEs, and any additional resources needed. This action has been completed by PHMSA, and a report was submitted to Congress on December 20, 2012.

Section 32—Authorization of Appropriations

This section of the act required PHMSA to ensure at least 30 percent of the costs of program-wide Research and Development (R&D) activities are carried out using non-Federal sources. These efforts are currently ongoing and are on-track.

This section additionally mandates that PHMSA transmit a report to Congress on the status and results-to-date of implementation of the R&D program every 2 years. The R&D program is designed to identify gaps in needed pipeline technology and map a path forward to assure there is no duplicative research and that resources are leveraged appropriately. PHMSA is finalizing a draft of this report.

III. SISSONVILLE AND THE CHALLENGES WE FACE

Despite our successes, we continue to face challenges in fulfilling our mission, and this is obvious when taking a look at what happened in Sissonville, WV. The explosion at Sissonville, as Chairman Rockefeller has said, was terrible, serious, and dangerous. Although several homes were destroyed or damaged, and portions of a major interstate highway were severely damaged, it is fortunate that no one was killed and there were only minor injuries. It could have been a much larger tragedy. We are working closely with the National Transportation Safety Board (NTSB) and the Public Service Commission of West Virginia in the investigation, and we are also undertaking our own compliance investigation. In addition we are taking immediate action to determine what additional steps need to be taken to prevent accidents like this from occurring in the future.

We have issued a Corrective Action Order (CAO) based on our preliminary findings. The pipeline will not be placed back into service until we are completely satisfied with the restart plan that Columbia Gas is required to submit. When the pipeline is eventually placed back into service, it will operate at a 20 percent pressure reduction from the maximum allowable pressure, while our engineers oversee a series of tests and evaluations and review the results. It is only after PHMSA is fully satisfied that the pipeline is safe for full operation that the pipeline can return to regular operating pressure.

One of the greatest challenges that we as an organization face is assisting our State partners to succeed in the inspection, regulation, and enforcement of the pipelines for which they are responsible. With the exception of Alaska and Hawaii, State pipeline safety agencies are the first line of defense in protecting the American public, and they have always been a critical component of PHMSA's success.

Thanks to provisions in the Act, we are currently able to cover 77 percent, or approximately $43.5 million, of the program costs that States incur. This funding covers personnel and equipment needs, public outreach programs, and other activities that allow the States to inspect and regulate intrastate pipelines. Currently, we partner with 52 state pipeline safety programs through certification and agreements for the inspection of the nation's intrastate gas and hazardous liquid pipelines. PHMSA also has interstate agent agreements with 10 states to perform interstate pipeline inspections. We are pleased to report that the State of West Virginia participates as an interstate pipeline agent for gas transmission lines. This partnership has proven to be a great asset in helping to strengthen the safety of pipelines in West Virginian communities.

The day this incident happened, several of my top staff members and I were visiting the Marcellus Shale area. We received a call that alerted us to the incident, and we were able to launch our response from the meeting we were conducting in Pennsylvania. Tim Butters, my Deputy Administrator, was in contact with emergency response officials from Sissonville shortly after the explosion occurred. It is because of the great relationship PHMSA and our State partners have with the pipeline industry and emergency responder community that we were contacted directly for support. PHMSA exists for the safety of the public, and we have been involved from the onset of this incident up through this point in time. We continue to support our fellow partners on the ground at the incident. As well as work with the emergency response community in order to share best practices and lessons learned.

In fact, we recently returned to Sissonville to meet with the local emergency responders and emergency management officials of Sissonville and Kanawha County to discuss the response to this incident, and what prior interaction they had with the operator.

We were very encouraged to learn that there was a good working relationship with the utility operator and the local public safety community. These established relationships, coupled with the fact that the local responders were well-trained, made it possible for the successful and effective management of this incident. The fact that there were only minor civilian injuries and no injuries to emergency responders is a testament to the capability of the local emergency response system and the importance of cooperation with the pipeline industry, and Federal and state regulators.

However, we also learned there is still much work to do. Both the pipeline operators and local officials recognize that additional training and exercises are needed. As the statute now requires, operators will be providing more detailed information about their pipeline systems, including location, size of pipe, and other critical elements. A rulemaking is under consideration that will allow PHMSA to collect additional information as part of its emergency responder outreach program. While Columbia Gas had been engaged with the local community, we were informed that cooperation and coordination between the local community and other pipeline operators could be improved. We will do what is necessary to ensure that this is corrected as quickly as possible.

We always make an aggressive effort to apply the information from specific pipeline incidents to the broader, national context of pipeline safety. We accelerated the implementation of control room management regulations based upon lessons learned about supervisory control and data acquisition (SCADA) system challenges. This year we will hold a public workshop to evaluate lessons learned during the last ten years of performance based integrity management regulations.

Lessons we learn from the Sissonville incident will also be used to help prevent accidents in other communities and will help us continue to fulfill the safety goals and purpose of the Act. Once our investigations into this incident are complete, we will release our findings and information to the larger emergency responder community and operator network.

IV. Changing Infrastructure and the Importance of Oversight

Much like the members of this Committee, this Administration has recognized the need for an aggressive approach to the safety of the Nation's pipeline system and the Fiscal Year 2013 Budget includes a funding request to implement an aggressive Pipeline Safety Reform initiative, which seeks to significantly increase both Federal and State resources supporting pipeline safety, as well as furthering research and development, and enhancing information technology capabilities to address the safety of the national pipeline system. We just recently received the final GAO study on the ability of transmission pipeline facility operators to respond to a hazardous liquid or gas release. We are currently reviewing the findings and will be happy to discuss with your staff on how we plan to move forward.

From the discovery of vast energy shale deposits, which will require the creation of additional infrastructure, to the maintenance and rehabilitation of the infrastructure already in place, the Nation's infrastructure needs are growing and changing.

I have been to the Bakken and Marcellus Shales, and I have seen these changes and the evolution of the energy industry firsthand. And I can tell you that we must prepare for these new and shifting demands right now. We must make sure that people and the land are protected at the beginning of the process even before the pipe goes in the ground. Effective standards and regulations are one of the best ways to keep America's people and environment safe while providing for the reliable transportation of the Nation's energy supplies, and the oversight provided by PHMSA and our partners will become even more critically important in the future.

With that being said, I believe that the Pipeline Safety Act, and our outreach and oversight, is working. We have a long way to go to reach our goal of no deaths, injuries, environmental and property damage, or transportation disruptions, but we have a solid foundation to build on as we continue to advance pipeline safety.

In closing, we look forward to continuing to work with Congress to address pipeline safety issues and to improve pipeline safety programs. Together, we will keep America's people and environment safe while providing for the reliable transportation of the Nation's energy supplies. Everyone at PHMSA is dedicated and committed to fulfilling the remaining mandates and accomplishing our pipeline safety mission. It is an honor to serve the American people and to work with the dedicated public servants at PHMSA. Thank you again for the opportunity to speak with you today. I would be pleased to answer any questions you may have.

In: Pipeline Safety
Editor: Maeve B. Reagan

ISBN: 978-1-62618-337-7

Chapter 4

TESTIMONY OF DEBORAH A. P. HERSMAN, CHAIRMAN, NATIONAL TRANSPORTATION SAFETY BOARD. HEARING ON "PIPELINE SAFETY: AN ON-THE-GROUND LOOK AT SAFEGUARDING THE PUBLIC"*

Chairman Rockefeller, Members of the Committee, and Senator Manchin, thank you for the opportunity to address you today concerning the National Transportation Safety Board's (NTSB) ongoing investigation of the pipeline rupture and fire in Sissonville, West Virginia, 7 weeks ago.

Mr. Chairman, as you have indicated, this is the fourth Senate Commerce Committee hearing on the issue of pipeline safety during your tenure as chairman. This hearing is also the NTSB's fourth Senate Commerce Committee hearing on this issue since I became Chairman. It is regrettable that major pipeline safety accidents continue to be a significant transportation and public safety concern. It is also regrettable that in the area of pipeline safety, philosopher George Santayana's aphorism that those who do not learn from history are doomed to repeat it, is certainly true. Indicative of the safety risks posed by pipelines, just four weeks prior to the Sissonville accident, the NTSB added pipeline safety to its Most Wanted List of the top 10 transportation safety challenges for 2013—the first time this general subject has appeared on our annual List.

* This is an edited, reformatted and augmented version of a testimony presented Jan 28, 2013 before the Senate Committee on Commerce, Science and Transportation.

Today, I will discuss the safety risks posed by the transportation of oil and natural gas by pipeline, the rupture and fire that occurred in Sissonville on December 11, 2012, the NTSB's response to the accident and the status of its investigation, and key NTSB findings and recommendations as the result of its past investigations of major pipeline accidents.

As described in our Most Wanted List, today, in the United States there are some 2.5 million miles of pipelines transporting natural gas, oil, and other hazardous liquids, with a significant amount of new pipeline design and construction activity underway. The pipeline network in this country includes 300,000 miles of gas transmission pipelines. Because pipelines are usually underground, most people don't even know they exist, much less where they are located. Therefore, it is incumbent on pipeline operators and regulators to ensure that the nation's pipelines are safe. Sufficient resources should be available to regulators to carry out critical oversight and enforcement efforts. These pipelines power thousands of homes and deliver important resources, such as oil and gasoline, to consumers. While one of the safest and most efficient means of transporting these commodities, there is an inherent risk that can lead to tragic consequences, especially when safety standards are not observed or implemented.

As was evident in Sissonville last December 11, pipeline ruptures can cause significant damage. Last July, the NTSB issued its accident report for the July 2010 hazardous liquid pipeline rupture in Marshall, Michigan—a rupture that was not discovered for over 17 hours. As a result, almost 850,000 gallons of crude oil spilled into the surrounding wetlands and flowed into local waterways, costing nearly a billion dollars to date for clean-up and recovery—by far the most expensive environmental clean-up for an onshore oil spill. Also, in September 2010, one of the worst gas pipeline ruptures occurred in San Bruno, California, when a natural gas transmission pipeline ruptured and ignited, killing 8 persons. In addition, 58 persons were injured, 38 homes were destroyed and 70 more were damaged as a result of this horrific and tragic accident.

The Sissonville Accident

On December 11, 2012, at about 12:41 pm eastern standard time, a buried 20-inch diameter natural gas transmission pipeline (Line SM-80), running west to east, perpendicular to Interstate 77, and owned and operated by Columbia Gas Transmission Corporation, ruptured about 112 feet west of Interstate 77 in

Sissonville, Kanawha County, West Virginia, near Route 21 and Derricks Creek. The pipeline maximum allowable operating pressure (MAOP) was 1,000 pounds per square inch gauge (psig), and the operating pressure at the time of the rupture was about 929 psig. After the escaping high-pressure natural gas ignited, fire damage extended nearly 1,100 feet along the pipeline and about 820 feet wide. About 20 feet of pipe was ejected from the underground pipeline and landed more than 40 feet from its original location.

The rupture occurred in a pipe that was a part of a pipeline segment installed in 1967 with a nominal wall thickness of 0.281 inches. The 20-foot ejected section of the pipe was found to have a fracture along the entire longitudinal direction at the bottom of the pipe. The outside surface of the pipe was heavily corroded near the midpoint and along the longitudinal fracture. The thinned area was approximately 6 feet in the longitudinal direction and 2 feet in the circumferential direction. The wall thickness had degraded so significantly that it measured only 0.078 inches at the point along the fracture—about 70 percent thinner than the uncorroded pipe.

The force of the released gas created a crater about 75 feet long by 35 feet wide and up to14 feet deep. Escaping high-pressure natural gas from the ruptured pipeline ignited. The intense fire destroyed three near-by homes, caused damage to several others, and heavily damaged both the northbound and southbound lanes of 1-77, closing both lanes for about 14-19 hours until the roadway surfaces were repaired.

The first call to 911 about the pipeline rupture and fire was made by a person at a nearby retirement home at 12:41 p.m. At 12:43 p.m. the Columbia Gas controller on duty at the gas control center in Charleston, West Virginia, began receiving alerts on the Supervisory Control and Data Acquisition (SCADA) system from instrumentation at the Lanham Compressor Station, located 4.7 miles upstream from the rupture location. Over the next ten minutes, 16 SCADA alerts indicated that the discharge pressure was dropping on Line SM-80 and two other pipelines in the SM-80 system (Line SM-86 and Line SM-86 Loop). The first notification to the Columbia Gas control center in Charleston, West Virginia, was provided by a controller from Cabot Oil and Gas Company at about 12:53 p.m., who had received a report of a "huge boom and flames shooting over the interstate" from a field technician who was near the accident location. Columbia Gas SCADA data indicate that the discharge pressures on the three pipelines leaving Lanham had dropped about 100 psig.

At about the same time that the control center was notified of the rupture, a Columbia Gas Operations Manager was called by a separate Columbia Gas field operator and told about the release and fire. The Operations Manager sent a crew to the Rocky Hollow valves approximately 3.2 miles downstream of the rupture, where two technicians, closer to the accident site, had already self-dispatched. Columbia Gas field technicians closed the downstream isolation valves at about 1:19 p.m., preventing the backflow of gas. The Operations Manager also notified personnel at the Lanham compressor station to shut the upstream valves. The 6 valves at the Lanham compressor station required a technician for closure. Technicians started closing the valves at 1:15 p.m., and notified the Operations Manager at 1:40 p.m. that the valves were fully closed, stopping gas flow to the rupture nearly one hour after the rupture occurred.

The NTSB's Investigation

After learning of the accident, a 10-person team from the NTSB, led by Board Member Robert Sumwalt, launched to Sissonville. According to our team's surveys conducted at the accident site, the rupture occurred in a nearly 38-foot long pipe joint that was a part of the pipeline segment installed in 1967. According to Columbia Gas documents, the ruptured segment of Line SM-80 was pressure tested twice in 1967: first at about 1,800 psig and then at about 1,750 psig. According to Columbia Gas records, the nominal wall thickness of the 20-inch ruptured pipe segment was 0.281 inches, had a longitudinal electric resistance weld seam, and was manufactured according to American Petroleum Institute specifications.

Parties to the Investigation are: Pipeline and Hazardous Materials Safety Administration, (PHMSA), Public Service Commission of West Virginia, Columbia Gas Transmission Corporation, Kanawha County Sheriff's Office, and West Virginia State Police South Charleston Detachment.

The NTSB issued a preliminary report on the Sissonville accident on January 16. Our investigative work, including metallurgical analysis of sections of the ruptured pipe at our laboratory in Washington, DC, is ongoing. Additional reports, analysis and a finding of probable cause will come later in the investigation.

Recurring Pipeline Safety Issues

Although it is premature for the NTSB to determine the cause of the Sissonville accident, issue findings, or draw conclusions, there are a number of recurring safety issues we have identified in previous pipeline accidents we have investigated that merit highlighting today. In particular, these safety issues include:

- Automatic and/or remote control shut-off valve installation
- Use of in-line inspection tools
- Integrity management program
- SCADA training

AUTOMATIC AND/OR REMOTE CONTROL SHUT-OFF VALVES

The NTSB has long been concerned about the lack of standards for rapid shutdown and the lack of requirements for automatic shutoff valves (ASV) or remote control valves (RCV) in high consequence areas (HCA) and class 3 and 4 areas. As far back as 1971, the NTSB recommended the development of standards for rapid shutdown of failed natural gas pipelines. In 1995, the NTSB recommended that the Research and Special Programs Administration—the predecessor agency of PHMSA—expedite requirements for installing automatic- or remote control valves on high-pressure pipelines in urban and environmentally sensitive areas to provide for rapid shutdown of failed pipeline segments. The current PHMSA integrity management regulation, which was promulgated in 2003, leaves the decision whether to install ASVs or RCVs in HCAs to the gas transmission operator.

In Sissonville, it took the operator approximately 58 minutes after the pipeline rupture and explosion to stop the gas flow by closing manual shutoff valves. Although the operator did not identify an HCA associated with the site of the Line SM-80 rupture, as the NTSB has pointed out in previous accidents involving pipelines located in an HCA, the availability of ASVs or RCVs is an important tool in containing the safety risks after a pipeline rupture.

Use of In-Line Inspection Tools

One of the 13 recommendations the NTSB made to PHMSA as a result of the San Bruno pipeline rupture and fire is to require all natural gas transmission pipelines be configured to accommodate in-line inspection (also known as internal inspection) tools with priority given to older pipelines. This recommendation was predicated on the NTSB's concern that in-line inspection is not possible in many of the nation's pipelines, which—because of the date of their installation—have been subjected to less scrutiny than more recently installed pipelines. As indicated earlier, the Sissonville rupture occurred in a pipeline segment installed in 1967. Due to construction limitations such as sharp bends and the presence of plug valves, many older natural gas transmission pipelines, including the ruptured segment in Sissonville, cannot accommodate modern in-line inspection tools without modifications.

In-line inspection tools travel through the pipeline to determine the nature and extent of any anomalies in the pipe. Another option for this type of testing is hydrostatic pressure testing that yields information about the integrity of the pipeline.

In the NTSB's judgment, the use of specialized in-line inspection tools that identify and evaluate damage caused by corrosion, dents, gouges, and circumferential and longitudinal cracks is a uniquely promising option. Unlike other assessment techniques, only in-line inspection can provide visualization of the pipeline integrity throughout the entire pipeline segment and, when performed periodically, can provide useful information about corrosion and crack growth. Although in-line inspection technology has detection limitations (generally a 90 percent probability that certain type of anomalies will be detected), the probability of detecting a crack may be improved with multiple runs, and it is nonetheless a more effective method for detecting unacceptable internal and external pipeline anomalies before a leak or rupture occurs.

Integrity Management System Assessments

The Pipeline Safety, Regulatory Certainty, and Job Creation Act of 2011, enacted a little more than one year ago, includes a provision requiring the Secretary of Transportation to evaluate whether integrity management system requirements first set forth in the Pipeline Inspection, Protection, Enforcement, and Safety Act of 2006 (the PIPES Act), should be expanded beyond HCAs and report the analysis findings to this Committee and the Committee on Transportation and Infrastructure, U.S. House of Representatives, by early next January. If the Secretary determines that integrity management system requirements should be expanded beyond HCAs, the Secretary must issue regulations to implement these requirements after a Congressional review period has elapsed.

Although the NTSB certainly welcomes the statutorily-required evaluation and recognizes that Columbia Gas and other operators of natural gas transmission pipeline facilities in non-HCAs are not required to establish integrity management programs that meet minimum performance standards established in PHMSA regulations, the NTSB views these programs as important business practices that these operators should consider for implementation. In our San Bruno, California and Marshall, Michigan, investigations, we determined the Pacific Gas and Electric Company and Enbridge Incorporated, respectively—both of whom must comply with PHMSA's integrity management program requirements—nonetheless had ineffective programs. Deficiencies identified by the NTSB included use of inappropriate inspection methods and tools and failures to detect pipeline defects.

The NTSB does, however, recognize that achieving a robust and effective integrity management program—whether mandated or voluntary—requires dedication, sustained effort, and resources.

SCADA Training

As indicated above, the Columbia Gas controller on duty received 16 "pressure-drop" alerts—but did not receive any "critical" alarms—on the SCADA system, before receiving notification from another pipeline operator. These alerts showed the discharge pressure dropping on Line SM-80 and the two other pipelines in the SM-80 system.

The NTSB has addressed SCADA training in a number of instances. In 2005, the NTSB conducted a study of SCADA in liquid pipelines. The study examined the role of SCADA systems in 13 hazardous liquid line accidents investigated between 1992 and 2004. In ten of the accidents cited by the study, there was a delay in recognizing the leak by the control center operators. As a result of one of the NTSB safety recommendations resulting from this study and requirements enacted in the PIPES Act, in December 2009, PHMSA promulgated its control room management rule for pipeline facilities in Title 49, Code of Federal Regulations, section 192.631.

In the Marshall, Michigan pipeline rupture, the NTSB determined that inadequate training of control center personnel allowed the rupture to remain undetected for 17 hours, including two startups of the pipeline. In the San Bruno, California accident, the NTSB found "that it was evident from the communications between the SCADA center staff, the dispatch center, and various other PG&E employees that the roles and responsibilities for dealing with such emergencies were poorly defined."

As part of its investigation in Sissonville, the NTSB is looking into the operator's control room operations, its SCADA system, and the capabilities and training of its control room staff.

CLOSING

Although the rupture and fire did not result in any fatalities or serious injuries, the Sissonville accident could easily have caused significant injuries and fatalities. Pipeline accidents that have occurred in San Bruno, California; Marshall, Michigan; Sissonville; and elsewhere are devastating to the affected communities. Particularly regrettable are the recurring frequency of these accidents and the resource constraints that hamper regulators' pipeline safety oversight.

This concludes my testimony and I would be happy to answer any questions you may have.

In: Pipeline Safety
Editor: Maeve B. Reagan

ISBN: 978-1-62618-337-7

Chapter 5

TESTIMONY OF RICK KESSLER, PRESIDENT, THE PIPELINE SAFETY TRUST. HEARING ON "PIPELINE SAFETY: AN ON-THE-GROUND LOOK AT SAFEGUARDING THE PUBLIC"*

Good morning, Chairman Rockefeller and members of the Committee. Thank you for inviting me to speak today on the important subject of pipeline safety. My name is Rick Kessler and I am testifying today in my purely voluntary, uncompensated role as the President of the Pipeline Safety Trust. My involvement and experience with pipeline safety stems from my years as one of the primary staff members on such issues in the House of Representatives and my subsequent work with the Pipeline Safety Trust.

The Pipeline Safety Trust came into being after a pipeline disaster over thirteen years ago - the 1999 Olympic Pipeline tragedy in Bellingham, Washington that left three young people dead, wiped out every living thing in a beautiful salmon stream, and caused millions of dollars of economic disruption. While prosecuting that incident the U.S. Justice Department was so aghast at the way the pipeline company had operated and maintained its pipeline, and equally aghast at the lack of oversight from federal regulators, that the Department asked the federal courts to set aside money from the settlement of that case to create the Pipeline Safety Trust as an independent national watchdog organization over both the industry and the regulators. We have worked hard to fulfill that vision ever since, but with continuing major

---

* This is an edited, reformatted and augmented version of a testimony presented Jan 28, 2013 before the Senate Committee on Commerce, Science and Transportation.

failures of pipelines, such as the one in Sissonville, West Virginia that brings us here today, we question whether our message is being heard.

Born from a tragedy in Bellingham, but also riding on the facts and emotion of other tragedies in places like Edison, New Jersey; Carlsbad, New Mexico; Walnut Creek, California and Carmichael, Mississippi, we have testified to Congress for years about the improvements needed in federal regulations to help prevent more such tragedies. For years we have talked about the need for more miles of pipelines to be inspected by smart pigs. We have pleaded for clear standards for leak detection, requirements for the placement of automated shut off valves, closing the loopholes that allow a growing mileage of pipelines to remain unregulated, and for better information to be available so innocent people will know if they live near a large pipeline and whether that pipeline is maintained and inspected in a way to ensure their safety.

So here we are again after the very recent failure of a pipeline in Sissonville which completely destroyed three homes, damaged other homes, caused extensive damage to an interstate highway, and once again terrorized a community. This recent failure falls too soon after a spate of significant failures over the past few years in Michigan, California, Pennsylvania, Montana, and Utah. Many of these failures had common themes and common solutions that could have prevented or at least minimized their impacts. We have been asking for action on these issues in previous hearings following previous tragedies for years now. Last year, Congress passed the Pipeline Safety, Regulatory Certainty, and Job Creation Act of 2011, which began to move the regulators and the pipeline industry in the right direction on some of these issues, but the speed of review, rule making, implementation and enforcement of the needed changes was not sufficient to prevent the tragedy in Sissonville. It is our sincere desire not to be back in front of this committee again in the future saying the same things after yet another tragedy.

The vision of the Pipeline Safety Trust is simple. We believe that communities should feel safe when pipelines run through them, and trust that their government is proactively working to prevent pipeline hazards. We believe that local communities who have the most to lose if a pipeline fails should be included in discussions of how best to prevent pipeline failures. And we believe that only when trusted partnerships between pipeline companies, government, communities, and safety advocates are formed, will pipelines truly be safer.

Clearly trust in pipeline safety has now been lost in the community around Sissonville, so add those people to people in Michigan, California,

Pennsylvania, Montana, Utah and elsewhere, where people now question whether the industry, regulators and legislators are really doing all they can to keep people and the environment safe.

In my testimony today I will focus on areas that are pertinent to natural gas transmission pipelines like the one that failed in Sissonville. Since much of the pertinent information about the Sissonville failure, such as whether or not it had been previously inspected, what type of inspection was used, whether the failure site was within a high consequence designation, and the type of valves upstream and downstream of the rupture site, has not yet been released, specific conclusions related to this failure would be premature. I will also review areas addressed by the Pipeline Safety, Regulatory Certainty, and Job Creation Act of 2011, and needed safety areas that bill failed to address. These are the issues I would like to speak to today:

- Response times to pipeline ruptures
- Expanding and clarifying integrity management requirements
- Inadequate federal and state resources
- Non-regulated and under-regulated Gathering Lines
- Poor facility response planning (hazardous liquids)
- Lack of clear jurisdiction for new pipeline approval and routing decisions
- Pipe replacement programs (cast iron, bare steel, faulty plastics)
- Quantifying natural gas leak significance
- Depth of cover at river crossings
- Diluted bitumen study constraints

Response times to pipeline ruptures – One of the critical issues related to any type of pipeline rupture is how quickly the pipeline operator can identify that a rupture has occurred and then act to shut the pipeline down to minimize any further effects of the pipeline failure. In a perfect world, built in leak/rupture detection systems would alert a pipeline controller of a rupture immediately and allow for the quickest response to shut down the pipeline. Unfortunately, as the final report - *Leak Detection Study – DTPH56-11-D-000001,*_which was recently provided to this Committee by PHMSA shows, for all leaks on natural gas transmission pipelines less than 16% are initially identified by the current leak detection systems. Even for the larger major releases that should be more easily identified with such systems less than 50% of these failures are initially identified by current leak detection systems. What this means is that someone other than the pipeline controller, such as local residents or emergency response personnel, or field employees with the pipeline company are the ones that initially identify the pipeline failure, and precious time is then lost as this failure identification is then relayed to the control room.

Once a failure has been identified, the pipeline operator still needs to be able to shut down the valves on either side of the failure site so the natural gas roaring into the local community is minimized as much as possible. In the case where the natural gas ignites, such as in Sissonville, the closure of these valves is what can halt the blowtorch effect on the neighborhood and allow emergency responders to access the area to do their jobs. The types of valves in these critical locations, and how far apart they are spaced, play an important role in how quickly the fuel will stop flowing into the community. The final report on automated valves - *Studies for the Requirements of Automatic and Remotely Controlled Shutoff Valves on Hazardous Liquids and Natural Gas Pipelines with Respect to Public and Environmental Safety* - that PHMSA recently provided this Committee provides the following cost effective strategy for reducing the consequences of natural gas pipeline failures such as occurred in Sissonville.

> "For natural gas pipelines, adding automatic closure capability to block valves in newly constructed or fully replaced pipeline facilities may be a cost effective strategy for mitigating potential fire consequences resulting from a release and subsequent ignition provided...
>
> The leak is detected and the appropriate ASVs and RCVs close completely so that the damaged pipeline segment is isolated within 10 minutes or less after the break, and fire fighting activities within the area of potentially severe damage can begin soon after the fire fighters arrive on the scene."

Unfortunately, as was seen in the recent Sissonville failure, and even more dramatically in the 2010 San Bruno tragedy, the leak detection systems combined with the associated valves were not capable of meeting the timeline in this cost effective consequence mitigation strategy. While these leak detection and valve issues have been talked about for years, current federal regulations do not require such automated valves, and it appears adequate leak detection systems for natural gas pipelines are many years off and will only be developed if adequate funding is provided for ongoing research and development. We join with the NTSB in calling for new regulations to require these automated valves at a minimum in all High Consequence Areas[1]. The

Pipeline Safety, Regulatory Certainty, and Job Creation Act of 2011 fell well short of these requirements by only requiring such valves for new or fully replaced pipelines. This shortcoming of the 2011 Act should be corrected to ensure that people living along existing natural gas transmission pipelines, such as in San Bruno and Sissonville, are afforded this additional protection also.

One other issue that Congress should keep a careful eye on relates to the development of a performance-based response time for companies to respond to and shut down pipelines in significant events such as Sissonville. The recent GAO report alludes to such a standard in its recommendations, which in part state:

> "evaluate whether to implement a performance-based framework for incident response times."

We certainly agree with GAO that the first step is to improve the incident response data available so such decisions can be made based on clear facts. In submittals to PHMSA on this issue, and at numerous public meetings, the Interstate Natural Gas Association of America (INGAA) has tried to create a starting point for such a standard response time discussion by repeating its findings and commitment of:

> "In populated areas, INGAA members have committed to having personnel on scene within one hour to coordinate with first responders and isolate failures.[2]"

As the recent valve study provided to you by PHMSA, and mentioned previously states, to effectively mitigate potential fire consequences from natural gas pipeline ruptures the failed pipeline segment needs to be isolated within 10 minutes. While it is true that a good deal of the damage from such pipeline failures occurs in the first minutes after failure, there is also clear evidence from places such as San Bruno and Edison that faster isolation of failed lines can reduce fire consequences and reduce the terror that citizens within the area experience. This often needlessly prolonged terror is rarely figured into the equations for such response times to shut down pipelines, but talk to anyone that lives through one of these events and you will realize that the terror has ongoing personal effects for years. Getting operators on site to isolate the ruptured site within an hour means that it will frequently be well over an hour before firefighters can safely enter the area. For firefighters waiting to get access to a potentially growing fire scene, and for those who live and work in the areas at risk, particularly hard to evacuate populations, that hour would be interminable. We do not believe one hour is a fast enough response time, and we urge Congress to keep a careful eye on this response time discussion.

Expanding and clarifying integrity management requirements – The Pipeline Safety Trust has testified at numerous Congressional hearings on the need to expand integrity management processes for hazardous liquid and gas transmission pipelines beyond the current limited requirements of High Consequence Areas. Integrity management programs have shown value by being responsible for the identification and repair of thousands of flaws in pipelines over the past decade. Unfortunately these programs are only required on around 44% of hazardous liquid pipelines and 7% of natural gas transmission pipelines. This leaves thousands of people in more rural areas without the clear safety benefits that integrity management programs provide.

We are thankful that in the Pipeline Safety, Regulatory Certainty, and Job Creation Act of 2011 Congress asked PHMSA to study the expansion of integrity management beyond High Consequence Areas, and we are also encouraged that PHMSA has already undertaken two significant Advanced Notices of Proposed Rulemakings to get this process started. Many progressive companies recognizing the value of integrity management programs have already moved to include all of their pipeline mileage under these programs, and the Interstate Natural Gas Association of America has publicly supported the expansion of integrity management to all miles of gas transmission pipelines.

While the Pipeline Safety Trust has been very supportive of the integrity management programs and would like to see them expanded, it is also clear that the program needs to be reevaluated to ensure that it is working as originally planned. There are a few areas within the integrity management programs that we believe need to be reassessed to ensure they are moving safety forward as intended. We understand that PHMSA is already preparing for a review and update of the integrity management program for transmission pipelines, and NTSB has also questioned whether regulators have clear evaluation metrics to effectively inspect and enforce such performance-based regulations. The most well publicized example of an issue that undermines proper integrity management related to the San Bruno tragedy where a lack of proper records led to incorrect assumptions about the type and quality of pipe in the ground. While much effort has been put into this record verification issue, there are other concerns with the integrity management program that still need to be addressed.

For example, also in the San Bruno tragedy, and perhaps in the recent Sissonville failure also, the use of Direct Assessment as a tool to inspect these large transmission pipelines has come into question. From the record of the development of the original integrity management program for natural gas transmission pipelines, it is clear that direct assessment was included as a way to appease the industry and help them avoid the large cost of retrofitting their pipelines so they could use the most up-to-date and effective internal inspection devices. Engineers from within regulatory agencies have shared concerns with us that the use of Direct Assessment is often done incorrectly, and is rarely as effective as the other approved integrity management inspection methods. We hope that a complete and thorough review of the use of Direct Assessment is undertaken soon, and that clearer criteria are developed for when and how it can be used. We support the NTSB recommendations that address this point by calling for hydrostatic pressure tests for all older pipe, and that all pipe be configured to accommodate inline inspection devices[3].

One further piece of the integrity management program that we think needs to be reviewed is the repair criteria. Pipelines that do not fall under the integrity management rules have a fairly conservative safety factor built into the design and operation, to account for the fact that once put in the ground there are no current requirements that they be inspected using the best inspection technologies. The repair criteria under the integrity management program reduce this safety factor because it was assumed that companies would be regularly inspecting their pipelines and would catch any problems before they reach a critical state. As seen in many failures in recent years this is a dangerous assumption, so we believe the repair criteria within the integrity management programs need to be reviewed and probably tightened to ensure a sufficient safety factor is maintained, since to date integrity management assumptions have not always been accurate.

We are concerned that PHMSA has not issued proposed rules on the Advanced Notices of Proposed Rulemakings (ANPRMs) to update both natural gas and hazardous liquid pipeline safety requirements. The Trust, industry, and other stakeholders spent many hours developing comments to respond to the ANPRMs on pipeline safety needs, especially in the area of integrity management. We hope Congress ensures that PHMSA acts in a timely manner on these important regulatory issues concerning integrity management.

Inadequate federal and state resources - For years the Pipeline Safety Trust has served on one of PHMSA's technical advisory committees, has helped with PHMSA workgroups on specific pipeline initiatives, and has had a great deal of interaction with PHMSA staff at all levels of the organization. All these interactions have confirmed our belief that this small agency is critical to pipeline safety, but is not as effective as it could be because of a lack of financial and personnel resources. The same issues also apply to state regulators who actually have more inspectors on the ground. For these reasons we support PHMSA's 2013 budget request[4], which would provide additional funding to support the needed increase in inspectors and analysts, an Accident Investigation Team, an increase in state funding, greater research and development, and the development of the much needed National Pipeline Information Exchange to help ensure adequate and accurate information is being collected to make good safety decisions. We hope this Committee, as the Senate committee that has the clear understanding of pipeline safety needs, will work with your colleagues to obtain this critical funding.

Non-regulated and under-regulated Gathering Lines – With the huge increase in natural gas production in states such as West Virginia and Pennsylvania, thousands of miles of under-regulated or completely unregulated gathering lines have recently been installed and more are on the way. No one really knows how many miles of gathering lines are out there or where they are located or how many have releases because up until recently no one ever tracked them. For example, the March 2012 GAO report[5] on unregulated gathering pipelines stated "out of the more than 200,000 estimated miles of natural gas gathering pipelines, PHMSA regulates roughly 20,000 miles." While in years past these gathering lines were smaller and lower pressure, many of the new gathering lines now being used in formations such as the Marcellus Shale are the same size and even higher pressure than the pipeline that failed in Sissonville. Yet unlike the Sissonville transmission pipeline, the majority of these gathering lines in rural areas, which may have riskier safety profiles than the Sissonville pipeline, are completely unregulated by the federal government.

For the most part the 20,000 miles of gathering lines that do fall under PHMSA regulations are the gathering lines that lie within more populated areas. Again many of these "regulated" gathering lines in these populated areas are the same size and pressure as the transmission pipelines that failed in San Bruno and Sissonville, yet are not afforded equal level of pipeline safety protection. For example a transmission pipeline running through a town would be required to undertake the important integrity management inspections to help ensure its safety, yet a gathering line that has the exact same risk profile running through that same town is currently not required to ever undertake any form of the important integrity management inspection and risk analysis.

While the development of various natural gas shale plays around the nation has arguably been a boon to our energy supplies and economy, because of this serious loophole in the pipeline regulations it has also increased the risk to thousands of people in these same areas. This is a loophole that needs to be closed as soon as possible before we have to gather for another hearing after a tragedy along one of these under-regulated or completely unregulated gathering pipelines.

Similarly, there are numerous unregulated hazardous liquid gathering lines with characteristics similar to regulated hazardous liquid lines. PHMSA needs to adequately regulate these gathering lines. Congress should consider elimination of the term "gathering" line for hazardous liquids. Doing so would ensure that all oil gathering lines are regulated, as the State of Alaska has done for its oil pipelines.

Poor facility response planning (hazardous liquids) – The NTSB in its report on the Marshall, Michigan spill of nearly a million gallons of oil into the Kalamazoo River made numerous recommendations targeted at improving facility response planning for hazardous liquid pipelines[6]. We support all of the NTSB recommendations and hope they will be acted upon as quickly as possible. As we have testified to this committee previously, the review and adoption of such response plans is a process that does not include the public. In fact PHMSA has argued that it is not required to follow any public processes, such as those under the National Environmental Policy Act, for the review of these plans. If the Enbridge pipeline spill in Marshall, Michigan and the BP Gulf tragedy have taught us nothing else it should have taught us that the industry and agencies could use all the help they can get to ensure such response plans will work in the case of a real emergency.

It is always our belief that greater transparency in all aspects of pipeline safety will lead to increased involvement, review and ultimately safety. There are many organizations, local and state government agencies, and academic institutions that have expertise and an interest in preventing the release of fuels to the environment. Greater transparency would help involve these entities and provide ideas from outside of the industry. The State of Washington has passed rules that when spill plans are submitted for approval the plans are required to be made publicly available, interested parties are notified, and there is a 30 day period for interested parties to comment on the contents of the proposed plan[7]. We urge Congress to require PHMSA to develop similar requirements for review and approval of spill response plans across the country, and that PHMSA's review and approval of facility response plans for new pipelines be an integral part of any environmental reviews required as part of the pipeline siting process.

To encourage greater public education and awareness regarding these response plans, Section 6 of the Pipeline Safety, Regulatory Certainty, and Job Creation Act of 2011 required PHMSA to "provide upon written request to a person a copy of the plan." In April of 2012, three months after the 2011 Act became law, the Pipeline Safety Trust requested a few of these facility response plans from PHMSA. We received an acknowledgement of our request within 2 weeks, but nine months later we are still waiting to receive the plans requested. In the State of Washington if we request such a facility response plan it is normally delivered to us on a CD within the week. While we certainly understand that PHMSA is understaffed, such long delays in filling information requests does little to accomplish the Congressional intent for public education and awareness, and makes us wonder how long others are waiting for information also.

Lack of clear jurisdiction for new pipeline approval and routing decisions – Nearly everyone agrees that the people living along the rights-of-way of the pipelines in this country can serve a very valuable function as the eyes and ears for pipeline safety along those routes. Unfortunately, too often the lack of any clear routing process and overly aggressive tactics by right-of-way agents sour the relationship before it even gets started, leaving too many property owners disgruntled and no longer willing to cooperate on safety issues.

For interstate natural gas transmission pipelines FERC provides a predictable siting process that provides communities potentially impacted by proposed pipelines valuable information about the proposal and ways to have their concerns heard and hopefully addressed. For all hazardous liquid pipelines, and for intrastate natural gas pipelines there is no such predictable process or information source. Some states have developed their own processes, while others have not, allowing smaller and smaller pieces of the decisions to fall on cities, counties and townships that often lack much knowledge regarding the issues associated with pipelines. This mish mash of routing authority often leads to a high degree of frustration from property owners and local governments who will be impacted by these decisions, and we suspect does not lead to the best routing decisions. Throw into the mix the often early threat of eminent domain and it is easy to see why these routing decisions too often become news stories about gymnasiums full of angry people that ultimately undermine trust in pipeline safety.

While the problem is clear and being repeated more frequently because of our new sources of gas and oil, we hope that Congress will use its investigative powers to commission a comprehensive study on this important issue to help find a solution. The study should at a minimum look at the shortfalls of the current system, compare the outcomes from the FERC process to the outcomes that fall outside of FERC authority, and consider which federal or state agencies are best equipped to help make these routing decisions for the various different types of pipelines. The study should also discuss any added benefits such cohesive route planning may produce in the form of lessening impacts by encouraging pipeline companies to better share infrastructure and rights-of-way, and in comprehensive environmental analysis allowing public review of potential alternatives.

Pipe replacement programs (cast iron, bare steel, faulty plastics) – Section 7 of the Pipeline Safety, Regulatory Certainty, and Job Creation Act of 2011 required the Secretary to conduct a survey every two years "to measure the progress that owners and operators of pipeline facilities have made in adopting and implementing their plans for the safe management and replacement of cast iron gas pipelines." After years of knowledge of the problems associated with this old cast iron pipe, and continued failures causing death and community destruction, this survey, which PHMSA has posted on their website, serves as a good way of shining a light on the operators who have taken this problem seriously and those who may not have. This was a great first step but could be expanded to be even more effective.

Cast iron pipe is not the only type of pipe in the ground that has clearly known deficiencies. There are some types of plastic pipe that also have been identified as in need of replacement, and older bare steel pipe that lacks the important protective coating of more modern pipe also poses a threat. These types of pipe should also be added to the survey to provide a measurable metric of how well pipeline companies are doing to address these potential problems.

While the Pipeline Safety Trust's main concern is the replacement of these types of problematic pipes for safety reasons, we also realize that paying for these replacement programs is a complicated equation. Many of the companies that have these pipes operate as regulated monopolies with a guaranteed rate of return, so the success of replacement programs often also lies with how state utility commissions approve rates for these replacement programs. We certainly support companies getting a fair return on safety investments, but the mechanisms to provide that return have to be carefully crafted to ensure the ratepayers are not paying for more than their fair share or for replacing things just to increase the rate of return with no real safety benefit.

Quantifying natural gas leak significance – With recent failures and deaths from leaking natural gas distribution systems the public has come to question the safety of the very common small leaks, which both regulators and industry acknowledge. New technology has also been developed that allows a person to drive through a neighborhood and see these small leaks all around. Recent information estimates that between 1.4% to 3.6% of all natural gas could be lost during transport, storage and distribution[8]. A 2009 article in the Pipeline & Gas Journal[9] regarding just the cast iron pipe portion of the pipeline network stated:

> A significant source of natural gas losses from distribution systems is cast iron distribution pipes. U.S. cast iron distribution mains are estimated to have leaked 9 billion cubic feet (Bcf) of natural gas in 2007. This equates to $150 million worth of gas, assuming the average U.S. distribution price in 2007, or $50 to $115 million if gas were valued between $3 and $7 per thousand cubic feet (Mcf).

We are surprised that more information has not been developed to clarify the quantity and significance of such leaks. Often such small leaks do not represent a safety hazard, but it only makes common sense that the loss of such a potentially large amount of gas is a significant waste of a non-renewable natural resource. Furthermore, methane (the main constituent of natural gas) has a far more potent negative effect on climate change than carbon dioxide, so the real quantity of natural gas leaking from these pipelines is important to understand along with what efforts to correct these leaks may be cost effective. We hope that Congress will ask for a study to better quantify these leaks, and discuss the impacts they have to safety, user rates, resource conservation, and climate change. Following such a study, Congress should consider requiring PHMSA to monitor and address significant natural gas leak problems from pipelines, compressor stations and storage.

Depth of cover at river crossings - –Section 28 of the Pipeline Safety, Regulatory Certainty, and Job Creation Act of 2011 requires the Secretary to "conduct a study of hazardous liquid pipeline incidents at crossings of inland bodies of water with a width of at least 100 feet from high water mark to high water mark to determine if the depth of cover over the buried pipeline was a factor in any accidental release of hazardous liquids." That study has been provided to this Committee, and concluded that depth of cover at river crossings was a factor in at least 16 incidents since 1991. A recent Wall Street

Journal article[10] provides a good overview of this problem along just one section of one river:

> "The U.S. Geological Survey found severe scour last year at 27 sites surveyed along the Missouri River from Kansas City to St. Louis, with the riverbed deepened in places by nine to 41 feet. Other unpublished USGS research found more severe scouring upstream.
>
> Of the 55 oil and gas pipelines that cross the Missouri—which runs 2,300 miles from Montana to St. Louis—at least 24 have sections that lie 10 feet or less beneath the riverbed, within the range of scour observed on the river, according to federal records obtained via a Freedom of Information Act request. During recent inspections, operators discovered at least two of those pipes, in Platte County, Mo. and near Boonville, Mo., were exposed but didn't break.
>
> Federal law requires operators to bury pipelines a minimum of four feet beneath waterways. Many river engineers say that standard is grossly inadequate. A congressional research report this year said the 4-foot minimum "appears to be insufficient to prevent riverbed pipeline exposure."

PHMSA already has a rulemaking in progress where they could address these findings. It is our hope that PHMSA in its rulemaking will develop clear standards that required companies, when geologically feasible, to use horizontal directional drilling (HDD) to place these pipelines at a depth under such river crossings to avoid future failures.

Depth of cover is not only an issue at such river crossings. Every year pipelines are struck and damaged, often leading to serious consequences, because of a lack of sufficient cover. Federal regulations require that hazardous liquid and gas transmission lines "must be installed with a minimum cover," but the regulations do not require that that level of cover be maintained. In some parts of the country normal erosion has led some pipelines to be at very shallow depth or even exposed, making them an easy target for plowing and various forms of excavation. While certainly excavators have a responsibility to call before they dig near such pipelines, the current depth of cover regulations need to be analyzed to determine if a change is warranted. An additional benefit of extending integrity management principles to more rural areas is that the assessment of foreseeable risks of third party damage to pipelines in agricultural areas from lack of cover will be made a necessary component of an adequate risk assessment by the operators, requiring them to undertake mitigative and preventative actions.

Diluted bitumen study constraints - Section 28 of the Pipeline Safety, Regulatory Certainty, and Job Creation Act of 2011 requires the Secretary to "complete a comprehensive review of hazardous liquid pipeline facility regulations to determine whether the regulations are sufficient to regulate pipeline facilities used for the transportation of diluted bitumen." PHMSA has contracted with the National Academy of Sciences for that review, which is due out next summer. Because of the high profile nature of the Keystone Pipeline proposed to carry this diluted bitumen, many people are already voicing concerns about the industry membership in the NAS review committee, as well as the fact that it appears the committee will not be doing any new research, just relying on existing information, a majority of which comes from industry.

The 2010 Enbridge spill of diluted bitumen into the Kalamazoo River in Michigan made clear that when diluted bitumen gets out of a pipeline it presents a difficult challenge to clean up because so much of it is prone to sinking. We had hoped that the diluted bitumen study that Congress required would be broad enough to also answer questions about the need for greater cleanup preparedness and technologies along pipelines that carry this unique material, but PHMSA's contract with NAS does not cover these problems. For these reasons we hope that Congress will pay careful attention when the report is released next summer, and ensure follow up of any questions left unanswered.

Thank you again for this opportunity to testify today. The Pipeline Safety Trust hopes you will closely consider the ideas and concerns we have raised today. If you have any questions about our testimony, the Trust would be pleased to answer them and, of course, we stand ready to work with you and your colleagues on improving this country's pipeline safety laws that are so important to ensuring the well-being of millions of Americans and the healthy environment that is their birthright.

End Notes

[1] NTSB recommendation P-11-011, 9/26/2011.

[2] Interstate Natural Gas Association of America, 11/2/11, comments on ANPRM for Safety of Gas Transmission Pipelines, Docket# PHMSA-2011-0023.

[3] NTSB recommendations P-11-014 & P-11-017, 9/26/2011.

[4] U.S. Department of Transportation, Budget Estimates, Fiscal Year 2013 http://phmsa.dot.gov/staticfiles/PHMSA/DownloadableFiles/FY%202013%20PHMSA%20BUDGET.pdf.

[5] GAO, Collecting Data and Sharing Information on Federally Unregulated Gathering Pipelines Could Help Enhance Safety, Report #GAO-12-388, March 2012.

[6] NTSB recommendations P-12-001, P-12-002, P-12-009, P-12-010, 7/25/2012.

[7] Washington Administrative Code 173-182-640.

[8] Robert W. Howarth Ã Renee Santoro ÃAnthony Ingraffea, 2011, Methane and the greenhouse-gas footprint of natural gas from shale formations - http://www.pseheal thyenergy.org/ data/Howarth_Climatic_Change_Shale_Methane1.pdf.

[9] Pipeline & Gas Journal, New Measurement Data Has Implications For Quantifying Natural Gas Losses From Cast Iron Distribution Mains, September 2009 Vol. 236 No. 9, Carey Bylin, Luigi Cassab, Adilson Cazarini, Danilo Ori, Don Robinson and Doug Sechler.

[10] Wall Street Journal, Floods Put Pipelines At Risk, Jack Nicas, December2, 2012.

INDEX

A

B

C

D

E

F

G

H

I

J

L

M

N

O

P

Q

R

S

T

U